ONLINE

a Guide for Broadcasters and Listeners

ONLINE Radio

© Copyright 2023
Paul Alexander Rusling
All Rights Reserved

The right of Paul Rusling to be identified as the author of this work has been asserted by him in accordance with the Copyright, Designs and Copyright Act, 1988. No part of this book may be reproduced in any form or by written, electronic or mechanical means, including by photocopying, recording, nor stored in any information retrieval system, without the written permission of the author or in accordance with the provisions of Copyright, Designs and Copyright Act, 1988. This book must not be circulated in any format without prior permission of the author.

Images of equipment and company logos are used to identify and publicise products and services and are either used with permission of their owners, or have been specially commissioned for this book, in which case the copyright to the image is retained. All copyrights are fully acknowledged.

There are links on some pages for the convenience of readers in finding more information on products needed for Online Radio; in some cases we may receive a small affiliate commission for any ensuing sales.

Although every precaution has been taken in preparation of this book, the publisher and author assume no responsibility for errors and omissions. Neither is any liability assumed or accepted for damages resulting from the use of the information contained herein.

First Edition.   November 2023

ISBN Softback: 978-1-900401-42-5
ISBN Kindle: 978-1-900401-43-2

Published by World of Radio, HU10 7TL, England

ONLINE Radio

To my wife Anne

for her unstinting help and support

and to those who pioneered Online Radio,
making it possible for the world to hear
unlimited entertainment and information

ONLINE Radio

ACKNOWLEDGEMENTS

Thanks to the many who have provided advice and suggestions for this book and to those who provided pictures of their products (see below).

I'm particularly indebted to several of the pioneers who developed digital audio, particularly the late Alec Reeves CBE and others without whom, we would have no digital audio. And without digital audio its certain that we could not have enjoyed such a blossoming world of Online Radio. A new breed of pioneers carried the torch for 'online'; those pioneers at Virgin Radio led by David Campbell who opened my eyes to the possibilities in the 1990s, including James Cridland who is still a radio visionary today.

Sincere thanks to those who inspired me (and urged me) to write this book and gave so freely of their time, providing invaluable information, particularly my friends at many radio stations too numerous to list. I've drawn unashamedly on help and advice from my friend Derrick Connolly who also foresaw the many possibilities for online radio 25 years ago.

I must also thank several other leading broadcast engineers, many of whom encouraged my enthusiasm for online radio, including Bill Dennay the BBCs Chief Engineer until 1993, to Tom Yingst, one of the true high-power pioneers of Continental, RCA and Harris, Quentin Howard at *Classic FM*, Alan Beech of *Commtronix* and Steve Conway during his time at *Apple*. They are just a few of those among us who shared ideas and notions of how we could make Online Radio the reality that it is today.

I've included html links on pages to many of the products needed for Online Radio for the reader's convenience; in some cases we may receive a small affiliate commission for any ensuing sales of these items.

Finally, to my publishing team at World of Radio, to all my radio friends for their kind advice and input. We know that the medium is not the message, but will this be one of the last books written with contributions, before we are all replaced by Artificial Intelligence?

My sincere thanks to you all

Paul Rusling
E: Paul@OnlineRadioBook.com

ONLINE Radio

CONTENTS

1	WHAT IS ONLINE RADIO?	9
2	A HISTORY OF RADIO	19
3	TECHNICAL	29
4	LEGAL & LICENSING	51
5	STUDIO EQUIPMENT	63
6	FORMATS	105
7	FINANCE	123
8	OPERATIONS	129
9	TRANSMISSION & DISTRIBUTION	141
10	ONLINE RADIO RECEIVERS	151
11	LUMINARIES & LEADERS	163
12	SUPPLIERS	173
13	GLOSSARY	178
14	THE FUTURE OF ONLINE RADIO	183
15	BIBLIOGRAPHY	186
16	INDEX	189
17	THE AUTHOR	192

ONLINE Radio

INTRODUCTION

Online Radio has rocketed in popularity recently, although it is now thirty years old. The reason for its mushrooming in popularity is largely the result of technical advances being made and the availability of ethernet on portable devices, such as smart speakers and mobile phones. As transmissions move to ever higher frequencies and with improved standards, more bandwidth becomes available, and at lower prices.

It really is totally amazing to see how much Online Radio has grown and so many people are now choosing this form of entertainment over others.

In its early years (the 1990s) Online Radio required complex adjustment of difficult to obtain computer programmes and a certain amount of knowledge by the listener. Consequently, online radio became the territory of just the really nerdy types, technical gurus and few obsessive anoraks.

Today, all that has changed as many radio stations offer Apps and other "easy tune" routines enabling them to attract listeners who are not so interested in fiddling about with the innards of a computer, just to get their favourite radio channel – they want a simple "Click and Play" arrangement. When that is presented on their mobile phone, they will choose that as their 'go to first" platform, to the detriment of traditional radio platforms such as the analogue bands of AM, FM, etc. Even DAB, that darling of the radio establishment, that was once touted as the future of radio.

Radio listening to the traditional 'over the air' platforms is plummeting among some demographic groups as they shift their 'live radio' listening to their mobiles, most often using an App (see page 158).

This book is designed to explain what ONLINE Radio is, how it differs from traditional forms of radio transmission and its likely place in the media firmament of the future. The book reviews the development of digital radio and explains how and why different types of transmission came to be used. In discussing the equipment and software needed it suggests typical examples and points the reader to links they can obtain more information.

ONLINE Radio

With any scientific, and particularly electronic subject, engineers often attempt to explain things with copious amounts of mathematics and complex theories. This book is however aimed at the layman and, while some background of certain decisions might foster a better understanding of the subject, it's beyond the scope of this work. Complex algebraic formulae and calculations have been summarised as far as possible, to make the subject easy to understand.

While ONLINE Radio began in the nineties, by Autumn 2023 there were over a hundred thousand stations audible via the internet, or 'online' as broadcasters now almost universally refer to it. Over half of all linear radio listening in the UK is to digital radio (of all types: DAB, Online, DTT, etc).

This book explains ONLINE radio, how it works and how it developed in the UK. It also describes what is needed to launch an ONLINE Radio station and some operational advice and suggestions to make it successful.

Brave broadcast pioneers are redeveloping radio with much looser controls, especially the newest sector, Online Radio. Stations are mushrooming at an increasing rate with new 'free' radio stations proving that it doesn't need millions of pounds or corporate controls to succeed.

Television is taking its lead from the latest radio innovations. Transmission went digital some years ago in most markets and made terrestrial delivery of multi-channels possible on Freeview. Now Sky is following suit with its Sky Glass televisions that don't need a set-top box nor a dish to access channels; they are all fed down the internet. It really is Online TV

Radio is a continuous live stream of programmes and it's that content that makes it, not the medium that carries the audio to the audience. Its future is being driven by the thousands of small operators, whether they are tuned to on traditional wavebands, or accessed via the internet.

We wish Online Radio every success and God's Speed.

The Online Radio Book has a new service:
Regular **up-to-date news** about Online Radio
Plus suggestions of fresh
Revenue Generation ideas
for online radio stations
More details:
OnlineRadioBook.com

ONLINE Radio

Chapter 1

What is Online Radio?

To answer the question, we must first define 'radio' and be clear what is meant by the term, as many people don't accept that certain types of wireless broadcasting are 'radio.'

Radio means many things to different people. To some it's any kind of small box that gives audio entertainment. In fact, to the broadcast industry, the word *radio* means "sound communication using radio waves" or "through the ether", as opposed to "**online radio**" which is generally accepted to be radio stations (also called channels) which broadcast 'down the line', i.e. via the internet, rather than using a large aerial to launch the signal via a portion of the spectrum as radio waves.

ONLINE Radio

What is radio?
That is difficult question. To some it's just the receiver, either in a bedroom, in the kitchen or in the car dashboard. Radio has always been quite portable, much more than other forms of entertainment such as the record player and its' no wonder that the domestic consumers think of just the receiver when you mention the word radio to them. But as the means of getting radio signals into the audience's ears have diversified into mobile phones, laptops, tablets and computers then the industry has come to regard it as a much broader term, meaning our precious radio industry.

As radio has spread to other transmission platforms so it has grown and encompassed other ways to communicate with audiences. The information can be sent in text form or picture too, but it's still radio. The medium has also become much more of a two-way street, with listeners more easily able to respond and contribute their thoughts and ideas. Some branches of radio have become interactive with the companies involved in streaming, such as Pandora, making up playlists from their listeners' preferences. These drive personalised music channels which are still radio and are curated, but they have no programme hosts or the other traditional accompaniments that make radio so personable and such a welcome companion.

What some perceived as the biggest threat to radio is happening; the possibility for just about anyone to become the broadcaster themselves. Radio will never again be a 'closed club' of the establishment, the rich and the powerful. Radio is for everyone, not just reception but for making the broadcasts too.

Radio broadcasting is over a hundred years old now, but Internet Radio began only thirty years ago. It's quickly grown into a billion-dollar industry and is still growing fast. New developments are making it an irreplaceable part of everyday lives for hundreds of millions of people around the world.

The term radio comes from an old Latin word, **radius,** which means two things:
      (a) the spoke of a wheel, or
      (b) a beam of light.
It's easy to see now how the word 'radio' began to be used in the 1920s for anything involving wireless waves.

ONLINE Radio

Radio broadcasting developed in the first two decades of the 20th century as an extra from wireless communications – the transmission of information by the use of wireless waves, which are formed by the interaction of two types of invisible force fields – electric and magnetic. The E and M fields.

These two fields (electric and magnetic) when combined will generate **electromagnetic radiation** whose waves travel outwards from the device combining them, usually the final component of a transmitter, an aerial.

The Latin word 'radial' is very appropriate, reinforced perhaps by the other part of its definition, 'a ray of light', as light waves are merely extremely high frequency electromagnetic waves, as is infra-red heat and all visible parts of the spectrum.

Radio Waves

Electro-magnetic waves emanate outwards from their source. In radio this is known as an antenna or an aerial, as they are usually aloft. If they were visible, a pure wave appears sinusoidal in shape. Fluctuations in amplitude, frequency and phase are imposed on the basic wave in a process called modulation which is done to carry the information. That information might be simple signalling, or voices or other information such as pictures.

The fields of radiating waves are alternating ones; i.e., their phase alters direction continuously. The waveform is often a complex shape, but at its simplest, is best be thought of as a sine wave:

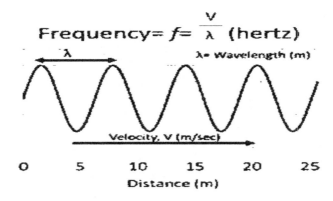

ONLINE Radio

Radio waves are measured by two main properties, that are interdependent:

Frequency: The number of times per second the wave makes a complete cycle, and
Wavelength: The distance (away from its source) travelled by the wave in one complete cycle.

If you know one property, its 'easy maths' to calculate the other, by diving either number into the 'Speed of Light', which as we know from the earlier comment is also the velocity of all electromagnetic waves, including radio waves.

Their relationship is $\dfrac{300,000}{\text{Frequency}}$ = Wavelength

and $\dfrac{300,000}{\text{Wavelength}}$ = Frequency

300,000 is the approximate speed of EM waves in metres / second, which is a velocity of 186,000 miles per second.

The simple radio waves described above do not have any information in them and are known as carrier waves. Information was originally added by switching them on and off quickly, which could be detected at the receiver. Morse code was already common and had been used for communicating simple messages down wires for over fifty years, using a code of short and long interruptions; dots and dashes.

After Marconi and others set up powerful stations to relay messages for others using wireless waves, such as to and from ships at sea, other experimenters found ways to add the human voice to radio waves and decode or decipher it back into audible sounds at the receiver.

Many advancements were made in World War 1 and afterwards so many people wanted to play with this new 'invention' that a cacophony of sound could be heard. International agreements were reached to divide up the spectrum into bands where some areas were reserved for communications (the various militaries snatched around half the available slots) and some for 'broadcasting' material designed for public reception, such as entertainment of news broadcasts.

ONLINE Radio

ONLINE RADIO
Online Radio has become the term for broadcasting radio stations, or channels, over the internet. The term "online radio" has largely replaced 'internet radio', but they are the same thing – it's just that the word 'online' has become predominant in use by radio stations 'on the air'.

Some argue that online this is "not proper radio", but that argument is flawed as it shows that they do not understand the term 'radio'. The word radio means simply the dissemination of programmes for general reception by multiple listeners.

The method of distribution is largely immaterial. In the case of 'online radio', the station's programme is sent along wires and fibre connections, which carry the internet. Online stations are also carried over some radio waves too along part of its journey, in the initial distribution legs of the journey, but usually for the final part of the journey too as most listening seems to be on equipment that is connects to WiFi for the internet feed, or via BlueTooth – and those are both simply terms that describe which part of the spectrum their radio waves operate in.

While traditional radio stations heard on a radio receiver (e.g. MW, FM or DAB) only cover a small part of the globe, online radio stations heard on the internet are, in theory, available in every part of the world, except those territories that limit access for political reasons, such as North Korea. Most people have the ability to "tune into" over 100,000 online radio stations

Online Radio is the branch of broadcasting that is growing most rapidly with the number of stations increasing almost daily, although some online broadcasters have fallen by the wayside, usually due to their lack of expertise and some over-ambitious plans being under-capitalised.

Receivers used for online look to proprietary directories to consolidate the station streams. These are best described as receiver aggregators or distributors, with the best known being *Delicast, Streamer, Radio.com, Tower, iTunes,* and *StreamingThe.Net*. *Tunein* has blocked a lot of global carriage following copyright battles. In 2021, the third-party aggregator platform *Reciva*, was closed by its owner, Qualcom causing several models of online radio to suddenly stop working.

Most online radio receivers use only one directory, which limits the number of stations they receive to about 50,000 but that's more than enough for most listeners. Carriage is dependent on the operator and on the online stations, which must propose their station for carriage on each network.

ONLINE Radio

Many of these replicate the stations carried on other networks, and in reality, stations only need to send an application to a quite small number of distributors.

Among the big advantages to online radio are that
- Stability, as reception not affected by radio waves or interference
- Often is commercial-free, so fewer interruptions
- Wider variety of programming with no regulators
- Extra links, such as purchase links, voting options, etc

Control of Broadcasting

While some governments allowed anyone to broadcast, within reason, other countries imposed strict regulations on broadcasts, limiting who could originate or even participate in broadcasting. In some countries, even the engineers had to be licensed, to ensure that reasonable technical standards were adhered to.

Some rebelled against this state control of radio and set up stations without a licence. The authorities branded them as 'pirates' and often hunted them down and seized their equipment. The most draconian of countries allowed no one but the state public broadcaster to transmit, in the UK this was the BBC. Originally a conglomerate of the big radio component suppliers (such as Marconi, Westinghouse, etc) the BBC had sole rights to broadcast radio programmes, with all other excluded and denied licences to transmit.

Some pioneering entrepreneurs set up companies and broadcast to the UK from transmitters on the continent. Most notable of these was the **International Broadcasting Company**, which built a station just across the channel in France, called *Radio Normandie*. Their programmes were very popular and it attracted higher audiences than even the BBC at times, especially Sundays when the BBC fasted on an output of hymns, sermons and religious talks. The IBC had its headquarters opposite Broadcasting House in London and a mobile studio which took its popular variety shows to towns around the country. At these, local artists were recorded onto large discs which were taken across the channel to the transmitter for transmission over the air.

The IBC bought airtime on other transmitters across the continent, including Radio Luxembourg. Despite it being a commercial success, it never resumed transmissions after World War II, although it did continue operating as IBC Studios until the 1970s and invested in offshore radio in the sixties.

TELEVISION

ONLINE Radio

After a somewhat shaky and confused start, the BBC's television transmitters were taken over in the war to help the development of radar. The world of television, like radio, remained limited to just one state broadcaster and all applications for independent licences were rebuffed. Costs were high in those years of austerity after WWII.

In 1955, a new government decided to allow limited commercial TV, and a highly regulated network of regional broadcasters were permitted from 1955 (ITV). Each station was assigned to one area exclusively, although there was some overlap in some parts of the UK, but coverage was never complete due to topography and poor drawing up of the original regions.

The major conurbations were carved up to be served by TWO companies; London for example had a service provided by *Associated-Rediffusion* from Monday to Friday with the weekends being served by ATV. They also ran ITV in the Midlands weekdays but at weekends, that area had programmes provided by ABC. The ABC studios also covered 'the north' at weekends, whose weekday programmes came from Granada TV (who also sold receivers on the high street!).

Many companies were joint ventures of two or more interested parties, with the theatre magnates of the Bernstein, the Delfont and Grade families dominating control of programming. They also trailblazed the way for selling UK-produced TV shows in many of the Anglophile markets around the world, such as the USA and many former British colonies and dominions, such as Canada, South Africa and Australia.

COMMERCIAL RADIO FOR THE UK

By 1960, Radio Luxembourg was the only radio station broadcasting to the UK commercially and then only after dark. From Easter 1964, Radio Caroline began transmissions using a ship anchored in international waters, neatly beyond the jurisdiction of the UK government. Many stations followed and almost a dozen ringed the UK's shores at one stage but half of these were under-resourced and adequate lacked funding.

ONLINE Radio

Attempts to starve-out offshore radio made life difficult for commercial radio broadcasters, most of whom disappeared after the Marine Offences Act made it illegal for British advertisers to use the ships. The demise of stations made life awful for listeners who now had only the BBC, until a small network of ILR stations launched in 1973. Restricted by tight regulation they failed to make much progress until the network was grown considerably and couple of national stations were licensed in the 1990s.

Only then did the UK's radio business begin to grow, both in terms of listenership and economic viability, although the revenues are still not as far advanced as in many other countries, despite the UK's commercial radio industry recently celebrating its golden anniversary. Many feel this is because most radio is in the hands of one private company and a German media publisher. In terms of programming, the UK punch above its weight, as the quality of British radio programmes abroad is very popular. The same is true for television and the TV industry thrives by selling UK productions to TV networks all over the world.

The birth of digital

Communications engineers first tinkered with the idea of 'slicing out' parts of the audible spectrum in the thirties which is when the British telephone and radio engineer **Alec Reeves CBE**, invented *Pulse Code Modulation*. Sadly, the technology was too cumbersome using valve technology and PCM had to wait until the invention of the transistor in the late 50s before it could come into regular use.

Reeves was a brilliant inventor with over eighty patents to his credit, including a radar circuit called *Oboe*, accurate to within 50 yards and that could not be jammed. He was also in the team that proposed using optical fibres for the transmission of audio, so Reeves did more than most in laying the groundwork for the development of today's Online Radio networks.

Kane Kramer is another British inventor and businessman to whom digital radio owes a huge debt, thanks to his invention of the first digital audio player, the IXI, in 1979.

It was **Carl Malmud** however who pioneered Online Radio, which he called "internet talk radio" in 1993. It was a weekly computer radio talk show, interviewing computer experts.
Internet or Online Radio?

ONLINE Radio

The only real difference between regular or traditional radio broadcasting and Internet or Online Radio is that the former uses the airwaves, or spectrum of electro-magnetic wave radiation to get the signal out to listeners and the latter uses the internet, or the web, to carry the programme. The word 'online' refers purely to the distribution method. Studios and programme formats are pretty similar for either method of distribution of transmission and are wholly interchangeable.

In the last two decades the studio equipment necessary to generate radio programmes has undergone remarkable transformation, as discussed below. The strategies in this eBook can work for every language and in every country because they are based on common denominators which can be found everywhere: human feelings, internet connection, communication and broadcast marketing.

Internet Radio is not an expensive medium and there are no barriers to anyone owning and operating their own Internet Radio station. You can spend a fortune on the very best equipment, software and other paraphernalia if you really want to, or you can launch your own station very modestly with the most basic equipment. You could probably start your own Internet Radio station with equipment you have in your home, though a couple of things would benefit from an upgrade; a decent microphone for example.

This guide attempts to explain what is needed, and why. It tells you where the necessary equipment can be obtained, and where it can be had cheaply. We've added many ideas and suggestions, but the book cannot be exhaustive and if you need more information on anything not covered by the book's main chapters, there are many extra resources in the appendices to this book.

Using a broadcast consultant to do some of the establishment could save a lot as they will know exactly how radio works, right through to transmission. Don't forget key questions of marketing, promotion of the station to listeners, as you can have the best radio station in the world but it's not going to be successful unless your listeners grow to become an audience. Is that audience marketable? No matter how niche it probably is, so marketing needs careful consideration. A broadcast consultant who knows the business and has been involved in it is probably vital to your success. We have suggested a few consultants in Chapter 11.

ONLINE Radio

Webcasting
The term for 'broadcasting over the internet', using streaming media, simply describes taking one source of content and broadcasting it to many different receivers simultaneously. Webcasts are usually broadcast live with a continuous stream of programme. If they are loaded onto a server for downloading 'on demand', this is called 'Podcasting', which has evolved into an industry of its own. "Radio on demand" is a good description. You can read more about Podcasting in this book, which was published in 2022.

https://worldofradio.co.uk/Podcasting.html

The term webcasting usually refers to linear streams or events which are non-interactive, i.e. the listener or viewer simply consumes the programme, and does not take part or participate in the programme. This is the traditional type of 'live radio' and it's called 'webcasting' only when it's not delivered via radio waves but by the internet.

Webinars
Some webcasting is not intended for large audiences, but is staged for a small limited and often restricted audience, usually using encryption or passwords to access it. This can be, for example, to internal staff groups spread over many sites for education, where it is often known as e-learning and also used for training sessions or staged for closed interest groups, such as diverse groups of investors or other stakeholders.

These 'limited appeal webcasts' might be interactive, with those watching also able to feedback, often by a simple chatbox. They are generally known as webinars, and may be encrypted or password protected. In situations where interaction is involved and those watching can offer feedback or participate, it's known as web conferencing.

SUMMARY
Online Radio is the distribution of radio programmes via the internet. It's also known as *internet radio, web radio, net radio* and *streaming radio*. Other names used for online radio are *e-radio* and *webcasting*.

Online Radio is one of the most common forms of streaming media, and involved presenting listeners with a continuous stream of audio that typically cannot be paused or replayed, much like traditional 'live radio'; it is distinct from on-demand file serving, or Podcasting, which is pre-recorded programmes that the listener can order up at any time – a sort of narrowcasting, accomplished singly.

Chapter 2

A History of Radio

Radio is now over a hundred years old and, while respecting that rich heritage, radio's future is of greater importance. When beginning any venture it help to know the history, successes and errors, to avoid repeating the same mistakes. Knowing the radio landscape helps too.

While TV is a key feature in most homes, vehicles and places of work, radio dominates lives in most countries. It is radio that still accounts for most media attention. Radio broadcasting is all linear audio delivery, whether achieved by EM waves in the ether or online; it's the content that pervades the audience's lives, not the medium that carries it. All audio broadcasting sent out on ether waves or by internet connections are all radio.

This is mainly due to the ability of radio listeners being able to do other things such as work, commuting, and other activities, while listening, which is a huge advantage over screen watching, be it TV or even simple texting. You can't work, walk and much less drive a car while watching a screen device, though many have tried, often with catastrophic results!

ONLINE Radio

When audio broadcasting began around 1920 radio was made mainly on long waves and soon after on the medium wave band. Stations often shared frequencies, which were uncontrolled with regulation in some places being only light.

In the UK and some European countries, the government kept a firm hand on broadcasting. Radio frequencies were regarded by the authorities as being far too important for a frivolous thing like broadcasting to ordinary people. This was shortly after the Great War and millions of men were still in military service. Electronics manufacturers were keen to promote sales of components and complete radios and many of the companies made experimental transmissions. This included the Marconi Company, Westinghouse, General Electric, Metropolitan Vickers and others.

The Marconi Company were the best known, as they actively developed the new idea of 'broadcasting'. Promotional events were common-place such as the world-famous soprano Dame Nelly Melba singing live at their transmitter in Essex. All broadcasting was live; the only recorded material was on phonographs.

By 1920, Marconi were one of the world's biggest manufacturers of communications equipment and had twenty years' experience in transmission. They allowed their staff to broadcast programmes of news, music, comedy and poetry to anyone who wished to 'listen in', as it was then called as this was felt likely to help promote sales of their equipment. London had two competing stations plus there were others in Birmingham and Manchester.

In some countries, programmes were sponsored by major newspapers or other commercial companies, but when formal applications were made for licences in the UK, the authorities rejected the idea. The GPO administered all communications and had complete authority over all post, newspapers, TV and radio.

The GPO regulated all communications in the UK and, in 1922, they insisted that ALL the equipment suppliers should group together and pool resources, forming the British Broadcasting Company. This was to be funded by the biggest equipment manufacturers and a levy would be added to the sale of every radio receiver. This encouraged more people to build their own receivers and radio construction became a popular hobby in the 1920s as a result.
Radio became so important by 1926 that the government ordered that the BBC, which was still a private company, should be handed over to a public

ONLINE Radio

corporation, subsidised exclusively by licence fees levied on its audience. The BBC has operated using that model ever since.

The BBC expanded to three services: the NATIONAL programme, several REGIONAL programmes and an EMPIRE service, which broadcast to the world on short wave. That later became the World Service and has now dropped 'British' from its identity.

Television

The world's first ever television programmes were made by John Logie Baird in 1923 who, within two years, was demonstrating television to the public in Selfridges store in London. He was soon transmitting the pictures over hundreds of miles and then across the Atlantic, using a part-mechanical method. Initially, pictures were transmitted on Medium Wave, with the sound on a second MW frequency.

By 1936 the BBC were sharing their transmitters between Baird and EMI, who offered a higher definition system which eventually was preferred. Baird then focussed on colour TV and had a system working by 1939 when WWII broke out and stopped radio and TV development in Europe. All experiments were focussed on the war effort, especially radar, the development of which led to dozens of small private radio manufacturers springing up.

The Birth of FM Radio

One of the biggest developments in broadcasting came from Edward Armstrong. He invented the regeneration and the superheterodyne circuits and then a system of adding modulation to radio waves by varying the frequency; FM was born! Previously, it had been claimed that FM was no better than AM, but by moving the frequencies ever high Armstrong was able to develop a wide-band FM which overcame the impulse noises that marred AM reception.

Armstrong conducted experiments in the RCA labs in the Empire State building but they were more interested in television so threw him out. Armstrong's demonstrations were well received by the press and by the FCC. Those first transmissions were in Band I, at 42 MHz but that was needed for television so, at the end of WWII, they were moved to Band II, 88 to 108 MHz, where they remain today.

In 1940, RCA realised that their dismissal of FM was a mistake and they offered Armstrong a million dollars for his patent, but he rejected their takeover bid and there began a bitter, lengthy legal battles over the new

ONLINE Radio

form of transmission. He later committed suicide in New York, following which his wife accepted RCA's offer and took the $1m, although the court battles continued until 1967.

After that, FM become more popular in the USA, although many restrictive laws, such as a non-duplication rule, curtailed real growth until around 1980. When first rolled out, FM had been the preserve of educational stations. The AM band was where the real action was, and the big profits.

Station operators would often turn to alternative outlets to use their FM allocations; the standard fare heard on the FM band was easy-listening or 'beautiful music.' often intended as background music for playing in elevators, etc.

Jazz and classical stations prevailed on the FM band, augmented by some foreign language stations, until well into the seventies. In 1967, Tom Donahue took over a daily four-hour slot on a small station in San Francisco called KPMX. It usually carried programmes in foreign languages and never figured in the ratings so Donahue began programming 'alternative rock' music, which was then called 'underground'. At this time, the American music business changing rapidly and San Francisco was the place where it was all happening! Within two months, KPMX had become all underground under Donahue's control and attracted huge *avant garde* audiences.

The KPMX DJs played psychedelic and experimental music that addressed the culture of sex and drugs more openly than any previous station. KMPX DJs adopted a mellow, laid-back style of presentation, as opposed to the 100mph pace of the AM Top 40 stations. The station played album tracks instead of hit singles, a format adopted by Radio Caroline in Europe in the 1970s.

Suddenly, the listeners were hearing albums that they'd never heard on the radio before. This was a world first as that kind of music had not previously been heard. KPMX went more commercial and limited what could be played, but the DJs rebelled and walked out. They all wound up on another FM station, KSFR which had an all-classical music format. Overnight this became THE voice of the underground and free speech, reborn as KSAN.

This new counter-cultural, hippie run outlet became the most important for the music for about ten years, and showed how to be a true community station, with free access to listeners who were invited to become involved in the station's output. Eventually, the commercial potential and some measure of political pressure overwhelmed the free-form, activist

ONLINE Radio

community stations. In spite of their popularity, the non-commercial segments began to disappear with even KSAN, the doyen of rock fans, becoming all country music for a while.

In the UK, FM broadcasts had begun in the 1950s, but the system didn't take off until the mid-1980s due to several factors. First, the BBC insisted on transmitting FM as horizontally polarised signals, meaning that portables and car radios got poor reception. They also kept their most popular station, Radio 1, off the FM band until the late 1980s, limiting its growth. It was a confined to an AM network shared with a megawatt station in Europe, marring its signal after dark. The BBC reluctantly made two hours of FM coverage available late at night, at a time when radio listening was at a low ebb and most music fans were tuned to Radio Luxembourg.

In the 1980s, dozens of unlicensed stations had begun using the FM band to broadcast a variety of formats to the capital. Some enterprising radio engineers found it was easy to install low power FM transmitters on tower blocks and cover most of the capital. A 'cat and mouse' game ensued with the most zealous radio investigators earnestly closing the stations down and seizing the equipment. Just as quickly, replacements appeared nearby.

The regulator of commercial radio stations was instructed to try new types of smaller stations and make provision of the most popular types of music that the 'pirates' were playing. These were mainly soul music formats, and so the tower block pirate stations such as *Kiss* and *Dread* were licensed as well as several non-English language stations aimed at the Greek, Turkish and Asian populations.

There were hundreds of applications for the new licences and the regulator moved very slowly. Applicants had to send in dozens of copies of multipage applications, with a high 'reading' fee and many other hoops to be jumped through, including assembling a board of directors packed with the 'great and the good'. This was well beyond the capabilities of many groups and certainly not small-scale radio! The smaller licences got bought out by ever larger groups, leaving today's situation with all the licences in the hands of just two or three operators.

ONLINE Radio

DAB RADIO

The standard called DAB began life as a research project in 1980. The IRT (*Institute fur Rundfunk Technik - the Institute for Broadcast Technology*) tested a system with engineers from the BBC, NRK and Deutsche Welle. After demonstrations to the ITU in 1985, the first broadcasts were made in 1988 as **Project Eureka 147**. The first transmissions in the UK were managed by the BBC and came from the Crystal Palace transmitter in South London. The first car radio with DAB cost £800 and the first DAB portables were horrendously expensive and very power-hungry.

Commercial DAB

The following year, the Radio Authority (the then regulator for all non-BBC radio) offered a licence to operate a single national commercial multiplex. There was only one applicant: the GWR Group who owned the only national commercial radio station, *Classic fM*. They joined forced with NTL form **DIGITAL ONE**, which came on the air in the last few weeks of the 20th century. It's first transmission was a recording of birdsong, made in the garden of their Chief Engineer, Quentin Howard. It had originally been made and used for the launch of INR-1, Classic *f*M in 1992. The birdsong channel continued for ten years and was also used for test transmissions of *Classic fM* in the Netherlands and in Finland.

In the early days a number of independent organisations operated national stations, but as DAB did not become an overnight commercial success these stations fell by the wayside. They were often replaced by "flavoured sub-brands" of stations from the large groups such as Bauer and Global, who wanted to fill the capacity of the multiplexes so that the DAB landscape did not start to look like a barren wasteland.

Small Scale DAB

Using DAB+ and cheaper transmitters expense of DAB encouraged some pioneers to explore lower cost methods. Their experiments forced OFCOM to licence over a hundred new mini-multiplexes (minimux) which are now launching in many small towns across the UK.

An unexpected result of the SS-DAB experiment's success is that some of the heritage stations are abandoning 'old hat' DAB, in favour of DAB+ and online coverage. Classic *f*M is among those making the switch. Many of their listeners are true HiFi afficionados, who are often the first advocates of Online Radio. They readily acknowledge that online gives the best radio reproduction as suffers from no interference or propagation artefacts.

ONLINE Radio

History of Online Radio

Before embarking on any venture it's important to know a little about history; both successes and the errors, to avoid repeating the same mistakes.

Internet broadcasting began in early 1993 when computer expert Carl Malamud launched his channel called *"Internet Talk Radio"*. Many stations experimented with the idea of relaying a feed of their regular output on the internet, taking advantage of the latest advances in digital compression. It wasn't long before software-based streaming players appeared, which made 'reception' of online radio possible on most computers.

In November 1994 a concert by the Rolling Stones was broadcast live on the web, with Mick Jagger launching the event by saying "I want to say a special welcome to everyone that's climbed into the internet tonight. I hope it doesn't all collapse!" That landmark concert, the first live one where an artist acknowledged the medium, marked the real public start of online radio, which is 30 years old in 2024!

Radio station **WXYC** in North Carolina is generally acknowledged to be the first radio station to broadcast on the web, using the *CU-SeeMe* software. This was originally a video conferencing programme, only available on Apple Macs. *RealAudio* was the first widely distributed streaming software, although it has since fallen in popularity. RealNetworks seems to be focussed on video distribution these days.

In Early 1966, ***Virgin Radio*** in London became the first European station to make its output available online and one of the first to be available 24 hours a day. Today its offering five channels online in the UK, some of which are simulcast on DAB, as well several dozen stations in countries around the world which are operate the *Virgin Radio* brand, with local station owners.

One of the longest lasting pioneers of internet radio was ***PureRockRadio*** (previously called *Radio 306*) in Saskatoon, Canada which ran from 1997 until 2021, before finally closing down.

In London, former pirate radio DJ Gilles Peterson is revered as for promoting international music on his online station **Worldwide FM** which was born out of a show he made for *Grand Theft Auto* in 2013. It's based at the BoatPod studio in Little Venice but has been operating restricted hours for most of 2023. *www.worldwidefm.net*

ONLINE Radio

The value of Online Radio

For a time it seemed that online radio was the new darling of media, especially when an IPO for new start up **Broadcast.com** saw its shares leap from the offer price of $18 shot up by 250% on the first day of trading, making it the first billion dollar 'dot com'.

The prospectus had admitted that they were not only losing money, but expected to do so for some time. Nevertheless, investors dived it and within nine months the company had been bought up by Yahoo for $5.7 billion!

The whole online radio business was almost killed off by American legislation, the Digital Millennium Copyright Act (DMCA). This made it mandatory for online and satellite stations to pay performance royalties, in addition to publishing rights. Traditional radio stations did not have any such imposition and considerable acrimony resulted, with pressure groups such as S**aveNetRadio** being set up to crusade for the penal payments to be rescinded.

SaveNetRadio was a coalition of listeners, artists, labels and online stations and, in 2007, it organised a 'Day of Silence' on which all online stations stopped playing music and ran protest messages. *Last FM* didn't take part, having just agreed a deal to be bought by CBS for $280 million; at today's prices after inflation that would be approaching $400 billion.

In the Summer of 2008, an enterprising Yorkshire-born engineer assembled a group of broadcasters to link a studio in Murcia in Spain to a server in Limerick in Ireland from where **Spectrum FM** could be heard globally on line, rather than the small chain of terrestrial transmitters it was using on the Costa Calida. Keith York and Steve Marshall's station attracted a loyal; following, combining high quality audio with carefully chosen, well processed, quality music.

The first really true network of online radio stations was formed when **NetRadio** was launched. A Minnesota-based company, founded by Scott Bourne and radio veteran Scot Combs, NetRadio was the world's first Internet-only radio network. NetRadio began using RealAudio in November 1995, transmitting just four streams of music with a licence from the American music rights organisation, ASCAP. At its height, **NetRadio** had 125 stations in its network thanks to its prominent position on *RealNetworks* software but, by 2001, it had folded.

In the late 1990s several radio and internet technology companies made public share offerings. *Broadcast.com* went onto the market with its IPO

ONLINE Radio

$18 shares soaring up to $68 on launch day. Things heated up quickly and within a year it was sold to *Yahoo* for almost $6 BILLION. Clearly, many saw Internet Radio as big business, not least the copyright owners and their agents, the collection agencies.

One of the earliest attacks on Internet or Online Radio was by the recorded music lobby. They persuaded the US Congress to pass legislation in 1998 forcing stations to pay performance royalties for playing commercial recordings. Traditional radio in the USA had always been free of 'performance royalties', although publishing rights must still be paid for. The result of the US Congress decision saw online stations paying far more than traditional radio, making a very uneven playing field.

That can be said to have held back the development of Internet Radio for many years. It kept Online Radio from the grasp of small independent operators who could not afford the high rates being levied, while media giants such as AOL could. The Digital Media Association claimed that even large media firms might fold or that many internet broadcasters would be moved to other countries where the royalty levies are not applied.

An organisation called *SaveNetRadio* fought a valiant battle to keep the costs down but eventually the Copyright Royalty Board imposed a minimum fee of $500 on every web radio station. Lengthy negotiations got the rates tied to turnover or even expenses for some classes of stations. This was called the Small Webcasters Agreement which was in force until 2016.

The first MP3 players came to market in 1998 but had only 32 MB of storage, so were more a toy than a commercial success. When Apple released the first iPod in October 23, 2001, it carried 1,000 songs and a 10-hour battery into a tiny case weighing only 6.5 ounces. A few years later the iPod mini could hold five times as many songs in a package just half the weight.

The increased functionalities that could be found on smartphones all served to make the iPod of less use to people. With the cost of internet falling the launch of music streaming services like Spotify and iTunes, were the rivals that "ate the iPod's lunch. The last of the iPods was released in 2022 and Apple say it is now obsolete technology.

ONLINE Radio

Chapter 3

TECHNICAL

In this book, we have deliberately avoided the overly complex mathematics of radio engineering; you won't need any algebraic skills, or knowledge of integration to understand the basic principles involved in DAB radio.

RADIO WAVES
Radio waves are type of electro-magnetic radio and are considered to radiate omni-directionally from their source (unless focussed into a beam, but that's a very complicated discussion). All electro-magnetic waves (from radio up to visible light) are considered to travel at the same speed, around 300,000 metres per second; that's close to the figure for a vacuum and it varies with the density and composition of the air they travel through.

ONLINE Radio

Electro-magnetic Spectrum

Extending from a few cycles all the way up to many Exahertz (10^{18}) also known as gamma waves, whose frequencies are seriously long numbers. One exahertz is one quintillion Hertz, that's a one with 18 zeros on the end! The numbers found most often in radio are only from Kilohertz to a few hundred Megahertz, which has only six zeros so is more manageable.

First, the units. A frequency can be expressed as cycles, which is simple to understand as it's a wave making one revolution. Cycles (and kilocycles, etc) are now known as HERTZ, after the German scientist Heinrich Hertz, the man who was the first to prove the existence of electromagnetic waves.

For many years, radio frequencies were names in cycles, or practice this meant kilocycles or megacycles, as the number of cycles was in the thousands or millions. Those terms have now been superseded by Hertz, thus we now have

KHz KiloHertz | **MHz** MegaHertz | **GHz** GigaHertz

It is important to have a good overview of where in the spectrum various bands are. Wavelength and frequency are often confused; each are a reciprocal function of the other. The wavelength (called by the Greek letter λ - *lambda*) is the distance a wave travels before it repeats the cycle, while frequency is the rate of repetition. Divide either into the speed of electro-magnetic radiation (300,000) and you get the other, as described earlier.

The higher frequency bands (and thus shorter wavelength) are much broader and can accommodate more signals, so this is where HiFi sound broadcasting was moved, to the VHF bands. There was ample space to accommodate the very wide band of frequencies needed. Television too needs a lot more space, several megahertz wide an d these signals were more easily accommodated up in VHDF and, by the 1960s, in UHF.

Mobile telephony too could be easily accommodated in the UKF bands and higher still. There is however one drawback – the higher frequencies do not travel as far as lower ones, for a given power level. They do however have shorter wavelengths, meaning that the resonant antennas are much smaller the higher the frequency gets.

ONLINE Radio

Frequency Bands

The properties of radio waves vary according to their frequency, and groups of them are arranged into bands, according to how the spectrum is divided up between the various users – broadcast, mobile, astronomy, shipping, etc. The best known in broadcast radio terms are Medium Wave, Short Wave and the VHF broadcast Band II (88 to 108 MHz) and a new one – VHF Band III, once used for TV but now the home of DAB.

This table is simplified and omits many other spectrum users, such as communications, beacons, amateur users, etc. The frequencies are typical and do not show the limits of bands.

Frequency (Hertz)	Band	Use	Wavelength (metres)
20 KHz	Extremely Low F	Global comms	15,000 m
100 KHz	Super Low F	Continental comms	3000 m
200 KHz	Long Wave	Continental Broadcast	1500 m
1000 KHz	Medium Wave	National Broadcast	300 m
2000 KHz	Short Wave	International	150 m
30 MHz	Short Wave	Broadcast	10 m
50 MHz	VHF Band 1	TV	6 m
100 MHz	VHF Band II	FM	3 m
200 MHz	**VHF Band III.**	**DAB**	**1.5 m**
750 MHz	UHF Bands IV & V	Television & mobile phones	40 cm
1.5 GHz	**L Band**	**DAB**	**20 cm**
2 GHz	S BAND	Sirius	10 cm
2.4 GHz	S Band	WiFi LANs	12 cm
30 GHz	Microwaves	Communications links	1 cm
430 TeraHz	Visible light	Red	700 nm
800 TeraHz	Visible light	Violet	380 nm
30 PetaHz	Ultra Violet Light		100 nm
300 PetaHz	X Rays	Medical & imaging	5 nm
300 ExaHz	Gamma Rays	Sterilisation	100 pm

On any frequency throughout the spectrum, it is possible to add any type of modulation, either analogue or digital, or any variation of analogue or digital, for example, FM, AM, DAB, DRM, etc.

ONLINE Radio

Channels & Frequencies

The frequencies used for radio broadcasting vary with geography and not every band is used globally. Given the effects of topography on radio signals, this is to be expected, to some extent, however some of the reasons have been political and done to protect markets. The soviet-aligned countries preferred to put domestic broadcasts on frequencies not used by the west. The most glaring example is their avoidance of the 88-108 FM Band II band in VHF. Soviet markets placed FM sound broadcasting outside the band used by the west (a few megahertz lower), to make it difficult for their subjects to tune to western radio stations.

Long, Medium and Short-wave band signals were regularly heard across many territories. Some countries made it an offence to have a short-wave receiver, but for incoming medium wave signals, all they could do was to jam out the 'offending' broadcasts from the west. Throughout the entire *soviet bloc* in the Cold War, each city would have its own transmitter site which broadcast strange noises over the bands to jam out western broadcasts.

The practice of jamming was largely discontinued by President Gorbachev during the glasnost and perestroika periods. After Russia became more democratic in the 1990s, the jammers all but disappeared, however in recent years more have been re-activated, particularly in the Russian republic, since its invasion of Ukraine and the ensuing war.

ONLINE Radio

Modulation

Modulation is the varying of a radio wave so that information can be carried on it, with either digital or analogue signals. Various forms of digital mod' can be used, and there are several ways of adding analogue information. Each method has advantages and some have disadvantages.

On the following pages we look closely at the various different types of modulation – starting with the oldest, Analogue types and then moving into the world of Digital, such as DAB.

ANALOGUE
Basically, analogue sounds are continuous waves of audio, or other information, such as our voices in the air, or musical instruments. This is sound at its purest – it's "full-fat" audio complete with a range of the basic (fundamental) sounds and harmonics.

An analogue waveform is a continuous signal that varies gradually over time. The basic analogue ways of varying a wave are to vary its Amplitude, its Frequency and also its Phase.

AM   AMPLITUDE MODULATION

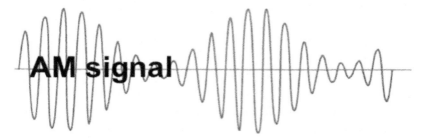

When radio broadcasting began in 1920, the programme was added to the carrier by varying the level of the signal up and down, in sympathy with the audio.

AM modulated signals can be easily demodulated and converted back into regular audio by the very simplest of circuits, including the legendary 'Cats Whisker' tuners as used at the birth of radio broadcasting, which have only three components. That simplest radio ever was improved by adding powered circuits using valves to boost audibility, enabling loudspeakers to be used.

ONLINE Radio

FM   FREQUENCY MODULATION

Just before WWII, several inventors developed a new system of modulation in which the actual frequency was varied in sympathy with the audio. Frequency Modulation was applied at a much lower level in the transmitter chain and, after field tests, was found to have several advantages.

The most useful of these was that it was not so prone to interference – particularly from pulses caused by car ignition systems and other nearby electrical equipment. As most interference is amplitude based, it is not difficult to use a simple limiter to reduce the noise, leaving only frequency variations, which can be demodulated leaving a 'cleaner' audio signal.

FM is not perfect however, having wide sidebands and also needing a more complex demodulator. The band used for FM broadcasting (around 100MHz) is prone to incoming interference from thousands of miles away with certain atmospheric conditions, such as high atmospheric pressure.

PM   PHASE MODULATION

Phase modulation is similar to frequency modulation (both are often called ANGLE modulation) and is an important technique in the transmission of data. It is similar to FM; both PM and FM can be considered derivatives of angular modulation however, a discussion of this involves complex mathematics and theory, which is beyond the scope of this book. In PM, the phase of the transmitter's carrier is varied in sympathy with the audio (or data). *Frequency Shift Keying* is a commonly used form of phase modulation which has been in use for many decades.

ONLINE Radio

Digital Radio?

The term DIGITAL Radio is often a misnomer as it is often used to refer to several different platforms. It is common to hear consumers call a radio 'digital' simply because it has numbers appearing on a screen display to denote the frequency (or sometimes even just the current time!). Those receivers are most likely NOT digital at all, but are analogue. Let's deal with this important difference immediately:

The human voice and all naturally occurring sounds are heard via an analogue acoustic wave created by vibrations. The human voice comes from a couple of vocal cords in the larynx, situated at the top of the windpipe. The vibrations pass through the air as an acoustic wave. They do not normally travel far (imagine the cacophony we would hear if they were – millions of voices and other sounds all being heard at once!) unless they are funnelled by acoustic pipes to be heard up to a few hundred yards away.

To get the human voice heard over a larger distance, they are either recorded or simply transmitted electronically to make them heard more widely. That involves converting the voice's acoustic sound wave into a varying electronic signal.

Sound waves are analogue in nature at origin, distinguishable by their frequency and their harmonics. Human voices vary from about 100 cycles up to 17KHz, (1KHz is the same as a thousand cycles) with the male voices pitched lower than females and not usually having much content above 8KHz. Most of the "intelligence" is contained in the band from about 350 to 3,000 cycles (human voices). This is the reason that telephone and, in particular, many communications circuits, are designed to focus on that narrow band of sound; adding in high frequency notes adds to the sound quality but at greater cost.

Early recording apparatus, microphones and loudspeakers were often very selective in the range of frequencies reproduce. Early microphones were based on carbon granules, whose resistance varied with the intensity of the sound, and the sound could be quite "tinny" and distorted.

The other big problem that constantly engaged broadcast engineers was how to get the electronic sound signals between studio centres and transmitters. Until very recently this was always done with special lines, often called "music lines" which had extended quality up to perhaps 10KHz, or even more, which was about three times that found on normal telephone lines. Also, the lines needed to have uniform characteristics; such aspects

ONLINE Radio

as frequency response could mar the audio fidelity if the capacitance or inductance varied much.

The lines should also be screened to prevent crosstalk from adjacent circuits also "bleeding through" into the programme line. This was dealt by with by having selective amplification along the line, or by screening the lines with an earthed jacket. Eventually, it was discovered that using a "balanced pair" of cables worked well up to quite high frequencies.

Recording Audio

The first recording machines were mechanical with acoustic waves being picked up on a diaphragm, onto which was connected a stylus. The waves vibrated the diaphragm, and caused the stylus to vibrate in sympathy. This was dragged through a narrow groove on a wax cylinder, or one with a thin layer of tin-foil. Rotating it formed a long groove which could give up to two minutes recording or playback time at a speed around 150 rpm.

cc

The audio could then be replayed by having another stylus track the groove, whose undulations caused a diaphragm to vibrate in sympathy with the groove undulations. It moved the air, enabling the audio to be heard once again. Fidelity was not wonderful.

The next step was to collect the audio in a microphone, which had better quality that then receiving horn used on the first phonographs. Its electrical variations would be connected to the stylus before cutting into the cylinder. Soon the wax cylinders were replaced by flat discs, which gave superior audio quality, and that basic system continues today with the discs made of vinyl and some of the most amazing quality can be heard on a system that is now enjoying quite a resurgence in popularity among audiophiles.

Machines using long think bands of steel were then used to record audio's electrical variations but they ran at alarming speeds and were dangerous if they snapped. They were replaced with safer spools of iron-impregnated mylar (strong polyester film) that could record up to an hour. Over the last half of the 20th century these 'tape recorders' could record with the purest quality and have become the benchmark for audio recording.

ONLINE Radio

Pure analogue audio places strict demands on equipment used to carry it and uses up large amounts of space if it's recorded without loss. Acoustic engineers have long known that a lot of the material contained of the acoustic wave is wasted, especially with the imperfections of the human hearing system and the way it degrades with age. A lot of the information can be discarded with little audible degradation of understand or perhaps even enjoyment.

Digitisation

Using very fast signal processing, any signal, audio or data, can be reduced by selectively chopping out large chunks of the signal that don't add much to its intelligence. This is called DIGITAL or digitising the signal. It is literally chopping it up into slices, which can then be transmitted more easily and cheaply. Another attraction is that it can be STORED more cheaply.

Digital Radio signals are not a continuous stream of information; the information is being rapidly interrupted, as with digital audio (e.g. CDs). It is more versatile than analogue as it can be more secure and is cheaper to transmit, although it is worse in quality and more easily degradable.

By slicing signals into a digital image of the analogue programme, many more audio channels can share one transmitter. The simultaneous use of one channel by two or more programmes is called **multiplexing**. This can be achieved in analogue, and many FM stations in the USA carrier extra channels in their subcarriers but is much more effectively done using digital modes of modulation.

Marconi developed a **PDM** (Pulse Duration Modulation) system in the 1970s for powerful AM transmitters. This was known as PULSAM, one of the first steps towards digital. PDM and **Delta Modulation** are simply ways of converting an audio signal into digital. Decoding a PDM signal back into analogue is a straight forward process.

By sharing, or multiplexing, signals more efficient use is made of the spectrum allocated. DAB signals are coded individually in the transmitter, before being multiplexed with other programme services and data. The multiplexer output is transposed to the appropriate radio frequency band, amplified and transmitted by antennas to the coverage area.

At the receiver, the appropriate carrier frequency is tuned in. The receiver selects this carrier (and its accompanying sidebands) and the output is fed into an 'OFDM demodulator' and channel decoder, which eliminates

ONLINE Radio

transmission errors. The output information is then passed to the receiver's audio sections and amplified to drive loudspeakers, playing the audio of the programme.

Frequency Division Multiplex (usually called FDM) is a way of spreading a signal over multiple, typically several thousand, closely spaced carrier frequencies, such that if one frequency is subject to interference, the signal can be successfully recovered and reconstructed from the remaining carriers. It is most commonly used for DAB, DRM and Digital TV systems.

A **DAW** or **Digital Audio Workstation** is a suite of music production software that lets you record audio and produce music on a computer. The first Digital Audio Workstation was the *Soundstream DAW* in 1977, combining a computer, disk-drive, VDU and the software needed to run it. Today's DAWs can record instruments as well as regular audio from microphones, and then loop, sample and mix to create amazing effects. It's essentially a complete audio production studio.

Today's DAW offers are simply software packages, enabling the user to record digitally, using most computers. The software has made audio creation and manipulation more convenient and easier than ever. The software is also getting more powerful.

Among the best DAW software options are:

- Logic Pro X
- GarageBand
- Reaper
- Avid Pro Tools
- Ambleton Live
- Audacity
- Studio One
- Bitwig Studio

ONLINE Radio

Digital Transmission

In digital transmission, the signal is switched on and off many times in quick succession, creating a series of zeros and ones, which represent either on or off. By using an optimum coding method of converting the audio, video or data into a digital signal, considerable efficiencies can be made, especially by discarding information not needed for intelligence.

There are many advantages and potential benefits to digital radio but for many years these were hindered by a lack of global agreement on standards.

Digital transmission requires less power than analogue, so it is more cost-effective and environmentally friendly.

There are many different forms of digital transmission, and different levels of encoding and compressing audio, drawn up by the *Moving Picture Experts Group*. The development of the MP3 format was done mainly by the *Fraunhofer Society*, a global network of scientists, whose UK base is at the Strathclyde University.

MPEG are a working group of authorities representing the fields of audio, TV, radio and video. They developed various standards for audio, which are known as MP2, MP3, AAC. There are others, such as Orbis, FLAC, Vorbis, etc. each of the audio coding formats has a standard or specification and it is known as an **audio codec**.

In-band on-channel, usually called **IBOC,** is a hybrid process of transmitting both analogue and digital signals together on a single frequency. Extra digital information (data) is added to a transmitter's sidebands with the analogue transmissions continuing as normal. This needs wider bandwidth, so has been adopted in the USA where the FM channels are 200kHz wide whereas most other territories space stations 100 kHz apart. (see below for operational details of HDRadio).

ONLINE Radio

Digital Audio Broadcasting has now been under development for forty years. As such, it is a mature format and some claim it's now obsolete having been surpassed and superseded by DAB+. The European patents for the DAB standard expired in 2013 and it's now a 'free-to-integrate' technology. DAB is incompatible with analogue transmissions, so uses different frequencies.

It's not possible to use the DAB standards in other bands, such as Medium Wave or short wave, which offer wider coverage. DAB is limited to short-range coverage and needs hundreds of transmitters. DAB's cost was of concern, though some improvements have been made. Even in the UK the regulator resorted to forcing stations to use DAB.

DEMISE OF DAB
The DAB standard was upgraded in 2006 DAB+ by applying the AAC coding standard, though the UK resolutely stuck to the old wasteful method, to avoid upsetting listeners who had bought early radios. AAC gives better sound quality than MP3 and is much better than the MP2 used in DAB. The DAB+ standard is more efficient, taking up less 'space', enabling multiplexes to carry more stations.

GROWTH OF ONLINE RADIO
Towards the end of 2023, many stations with 'double distribution' (heard on at least two different platforms) were seeing DAB being overtaken by online listening. By Autumn 2023, over 41% of adults are using online radio with almost a quarter (i.e. 13 million) now using Smart Speakers to listen to the radio. According to the latest RAJAR figures, released in October 2023, ¾ of radio listening is now via a digital platform. This includes Online Radio, which now enjoys a weekly reach of almost 41% or a quarter of total share. Most major BBC services in 2023 saw DAB reaching less listeners than FM. For commercial stations, online listening is higher still.

Some British stations that could be received both online and on DAB have now dropped the latter and have made their services ONLINE ONLY. Popular 1960s station **Atlantis FM** is a good example: they made their main service 'online only' in 2023 and director Rob Day reports that they saw no downturn in advertising, but are saving over £100,00 a year in transmission fees. So encouraged are Atlantis, they plan to launch three new 'online only' stations in 2024.

Online Radio is clearly proving to be Economically viable, if operated professionally.

ONLINE Radio

DRM ~ Digital Radio Mondiale.

A set of standards specifically designed to work with and alongside existing analogue transmissions on all bands, including AM, Short Wave and the FM broadcast band at 88 to 108MHz. DRM uses the spectrum allocated to it far more efficiently than AM or FM and uses an AAC based coding format. In tests in 2020, DRM was found to have the lowest carbon footprint, for transmission and reception.

Less than half the usual power need is typical, which can represent huge savings for most large broadcasters. This is for the same coverage with far better audio fidelity. Two manufacturers have announced a range of new DRM radios for launch in 2024.

DRM for mediumwave has been tested and documented about two decades ago. It has been recommended by ITU in 2005 and all the tests carried out all over the world are available openly and freely for anyone to see. Over three dozen stations in now use DRM covering 800 million people and seven manufacturers have radios in the market. Pakistan announced in 2023 it is moving its services over to DRM, bringing about 40% savings in power requirements. Pakistan sees the use of DRM essential in its new one million-watt (1MW) station now being built that will be audible from the Mediterranean to the Pacific.

Over a quarter of listeners to the BBC World Service still use AM and it is expected that the addition of DRM service should achieve good penetration in the volatile Levant region. The main deterrent seems to be only the availability of suitable receivers. In Russia, China and many other countries, the demand for Short Wave is still healthy and the recent wars in Ukraine, the Yemen, and other trouble spots, such as the Sudan, Taiwan and Korea, have resulted in a resurgence in the use of Short Wave, many of which now carry a DRM feed.

At the IBC in Amsterdam in September 2023, China announced that it too will be changing its domestic AM networks over to DRM, bringing the system to a further potential 1.4 billion listeners. Indonesia also announced that it is also making the switch, serving over 275 million listeners.

ONLINE Radio

HDRadio is a proprietary brand of "in-band, on-channel" (IBOC) digital radio that can be used on any traditional broadcast band, including medium wave or FM. *HDRadio* can use a single frequency to transmit up to six programmes. The system uses bit rates from 48 kbps on AM to 128 kbps on the FM band.

In many major American cities, HDRadio has become established in major American cities where car, home and portable radio receivers are readily available. Hundreds of car models now come with HD Radio installed and over 4,000 stations are on-air already. The FCC has given permission for all AM stations to go 'all digital' if they wish and some have done so.

When an HD Radio tuner loses the station's digital signal, it will automatically switch over to the analogue signal broadcast at the same frequency with just a slight break in sound. When back in range of a station's HD Radio signal, the radio reverts to it.

The new technology has been implemented in over 2,500 radio stations, covering over 400 million listeners in countries as diverse as the USA, Mexico, Canada and India. HD Radio technology was developed, and continues to be licensed by, iBiquity Digital Corporation. A major reason for the roll-out of HD radio technology is to offer some limited digital radio services while preserving the value of the existing radio stations.

DTT (Digital Terrestrial Television)
In late 2002, the Digital TV transmitters in the UK began carrying radio services on the spare capacity of Freeview. The BBC launched their new DAB-only channels plus *BBC Asian* and the *World Service.* In 2023 they added their main channels, simulcasting all AM & FM outlets. Several new commercial stations joined in and there has been quite some churn with many not lasting very long. All transmit 'free to air' but require a national licence from OFCOM broadcast on Freeview Play, which

UK Radio Portal
An innovative new service bringing Internet delivered radio channels to the UK's DTT Freeview network. It's accessible from the Freeview Electronic Programme Guide (EPG) on channel 277 on Smart TVs and many set-top boxes, including Sky's QBox and the Amazon Firestick devices. With over 20 million devices, it's one of the fastest growing platforms in the UK.

https://www.ukrp.tv

ONLINE Radio

DSR Digital Satellite Radio

Satellite has offered listeners radio channels on the sub-carriers of various channels since the 1980s. Many of these were carried to the UK on the Astra birds, each analogue channel could carry up to five radio stations. The receivers were not easy to set up and used a dish antenna, so were fixed installations, limiting the appeal and the take up. In Europe, both Eutelsat and the Astra carried hundreds of channels but mainly used for feeding remote transmitters.

The business of broadcasting by satellite never really took off in the UK due to several factors. The initial coverage charges by the satellite owners were very high. Secondly, the birds used were low power communications satellites and thirdly, to be geo-stationary, they had to be positioned above the equator, making signals in northern Europe quite weak, unlike those in North America which is mostly at lower latitudes.

((SiriusXM))

In the USA **XM Satellite** and **Sirius** each built satellites which offered radio signals as a primary service, and at a high-power level. The USA's lower latitudes mean that the satellite signals are stronger enabling the use of simpler antennas on.

They have now merged and use six satellites to give a service of almost 200 radio channels, largely available on subscription. Since buying Pandora in 2019, SiriusXM has continued to grow. They now provide music, sports, entertainment, comedy, talk, news, traffic and weather channels as well as podcasts and infotainment services to around 35 million paid subscribers. The most popular channels are music-led, with Howard Stern achieving more than a million listeners a week. The *SiriusXM* platform had a major revamp in November 2023 with a new logo, new channels, a new App and many in-car innovations.

Worldspace

International company Worldspace began broadcasting radio programmes to Europe, Africa and the Middle East in 1999. At its peak,
Worldspace attracted almost 200,000 subscribers but the company never achieved sufficient penetration to be commercially viable. Of the 62 channels it offered, two thirds were provided by independent suppliers such as *Radio Caroline* and remainder originated 'in house'.

ONLINE Radio

BIT Rates

The amount of data transmitted by any digital radio station has an effect on the audio quality that the listener hears, and it affects the amount of space. This makes it cheaper for stations to transmit at lower bit rates. Higher bitrates however make the signal more robust, less likely to drop-outs or interference and can impart more sound information, giving a fuller sound. Internet carriers limit the amount of data per month, just as DAB multiplex operators vary their carriage charges by the bitrate allocation.

Audio can consume a lot of data, so it is better to use the lowest bit rate tolerable. A 128k stream uses around 1MB of mobile data per minute, or 60MB per hour.

A 128 kbit/s signal is lower quality than FM, prompting complaints from audiophiles. By cutting that original 'standard' stream of 128kbps to only a third, i.e., 48kbps, as used in the DAB+ system, still enables an acceptable audio channel. It's about three times more efficient than DAB for a similar sounding audio.

The BBC transmits its primary national services as 'joint stereo' signal at 320 kbps on its *Sounds* online service. On DAB, it reduces most of these down to 128 kbps, or 192 in the case of Radio 3. The channels carrying speech are transmitted at 80 and even 64kbps.

The main commercial stations transmit at 64 and 128 online, while on DAB these are reduced to 24/32/40/64/80 kbps with Classic fM using 128kbps. Many of the commercial services are mono only. Many online stations offer lower bit-rates so listeners can keep their data usage down and save cost, especially on mobile phones.

Even by stepping down the quality and transmitting a mono signal at a lower rate, stations retain almost all their listeners, proving that people select a radio station for its programme, rather than audio fidelity. This is very similar to listeners choosing to listen to an AM station rather than FM, if AM offered the programme they wanted and FM didn't. Listener choice will always rule.

Some query the success of mono stations but it should be remembered that many listeners have only a one speaker radio, especially those listening on a smart speaker. Once again, content wins hands-down over audio quality, every time. All other things being equal, online radio is capable of far better audio quality than any radio carrier, including FM and DAB.

ONLINE Radio

Some of the fastest growing radio stations use very low bits rates. Union Jack is a local station that programmes only British Music (it launched in the wake of the Brexit referendum, in 2016) and it uses only a 32kbps mono signal. But managed to increase its listening by 73% in 2018. Some of the new commercial stations use mono streams at just 32kbps.

Many of the latest radio stations that are now meeting with substantial commercial success transit in mono only: *Absolute 90s, Heart Extra, Heat, Kiss, Planet Rock, Radio X* and *Smooth* for example. Clearly, the programme content is more important than audio fidelity.

Digital Audio Files

Audio is stored in various 'file formats'. All fall into one of three types: (a) lossy compression, (b) lossless and (c) uncompressed.

LOSSY COMPRESSION

MP3 also known as MPEG-1 Audio Level 3.

Windows Media a proprietary codec by Microsoft

OGG / Vorbis Free, better than MP3. Usage minimal.

AAC is a successor to MP3. Often called MPEG-4 AAC.

AAC+ is a development of AAC, aka as HE-AAC.

RealAudio a proprietary codec based on MPEG-4

LOSSLESS COMPRESSION
This does not lose any data which it does by packing data into a smaller file size by a code to signify redundant data. The main one used for audio is
FLAC Free Lossless Audio format (reduces size to 50 – 70%)

UNCOMPRESSED
The PCM files **WAV** and **AIFF** give best quality but are significantly larger than compressed file types. Both give similar quality, but Apple's AIFF can also contain meta data and is preferred by musicians and DJs. Technicians call both "full fat audio"

ONLINE Radio

There are several more digital audio formats, including some that are now part of the *Xiph.org Foundation*, which also has OGG Vorbis on its roster, such as **Speex, Opus** and **Theora**.

The amount of data that is discarded when an audio file is digitised varies between the format, but the size differences after processing for 'lossy' files is of the order of 10:1. Some 'Lossless' formats such as FLAC keep all the original material but use classic compression techniques to reduce the file size to just 2:1, which can still be a huge amount of data that's saved. The data is enclosed in an Ogg container, which makes it possible to add meta data (tags).

MPEG2 is the main digital audio format used in DVDs and on satellite transmission, which is now becoming dominated by H264. The main file types used for compressed images are JPEG, PNG, TIFF and Targa.

The main codecs used for transmission of video signals are: H264, MPEG2, VPS, AVI, MOV, RealVideo and Spark.

Other broadcast platforms

Traditional radio stations use platforms that developed over the last one hundred years using "over the airwaves' transmitters, on the bands of frequencies described above. The principal radio broadcast platforms are:

Long Wave Band
Using very low frequencies between 153 and 279 KHz. The Wavelengths corresponding to these frequencies are very large, from about 2,000 down to 1000 Metres. Accordingly, they have very large aerial masts which are often inefficient and so use very high power to launch the waves. The UK's most powerful radio transmitter is the BBC Radio 4 facility at Droitwich which is about 500 Kilowatts. Many European LW stations over the last fifty years have used powers up to 2,000 Kilowatts, but most have recently closed.

ONLINE Radio

Medium Wave Band
This is where most radio stations got their biggest audiences, in the days when radio was king. These wavelengths have been a life-saver for many, making it sad to see it now summarily dismissed. MW frequencies are those between 550 and 1600KHz, or the wavelengths of 200 to 550 metres. More modest wavelengths mean that aerials are shorter, but 100m high is still typical. Sometimes the tower itself radiates but often a mast supports a cage of steel cables which form the radiating element.

Using several masts of a few hundred metres high in a phased array means that considerable directivity can be obtained and the horizontal radiation pattern made to maximise signal in desired directions, or minimise in others, to protect other stations using the same frequency. Elaborate antenna farms with multiple towers are being made redundant as the demand for MW and LW broadcasting is made redundant by the growth in online broadcasting and other platforms. Medium Wave sites are usually located on the edges of large conurbations where there is a healthy demand for building land; high prices can be had by selling off these sites.

Short Wave Bands
Usually called HF (High Frequency) these transmissions are from 3 to 30 MHz, corresponding to wavelengths of 150m down to 10 m. Shortwave radio transmissions can travel thousands of miles, by bouncing off the ionosphere. This makes it possible to deliver content to millions of listeners around the world from a single transmitter. HF listening is decreasing as online radio replaces it, but many countries and organisations still spend millions every year, to get their signals into foreign countries.

VHF-FM
Even higher in frequency are the **VHF bands**, in most places the frequencies used are 88 to 108 MHz, are about 2M in length. Transmissions on these frequencies normally only travel just beyond the optical horizon, so the short aerials (just a few metres in length) are best located as high s possible, with many being stacked vertically to get greater directivity towards the horizon.

VHF isn't affected much by the ionosphere but any energy transmitted upwards is simply wasted as it is radiated out into space. Frequencies between 100 and 400 MHz can be ducted by the troposphere however, often carrying on for a thousand miles or so. This can cause problems to distant co-channel users, but is also a useful method of getting VHF communications traffic well beyond the normal range of the 'optical horizon'.

ONLINE Radio

The normal modulation used at the VHF Band II (88-108) is FM – **Frequency Modulation**. This takes up more space than AM, but the available space at VHF is much wider – 20MHz. The MW band by comparison is only 1 MHz wide, so in any given area, about 20 times more stations can be accommodated. Also, VHF signals travel not much beyond the 'optical horizon', so can be reused much closer, whereas the MW band is limited due to incoming interference from a wide area, especially after dark when the ionosphere bounces signals back down again.

The extra space at VHF means that more information can be added a signal, giving higher fidelity and stereo signals, which is much more difficult on lower frequency bands of MW, LW and SW.

A lot of research has been done into transmissions of digital signals on existing FM transmitters, particularly by the DRM proponents, but receiver supplies have not yet become available in sufficient quantities to make such a service commercially viable.

Digital Radio Mondiale (the last word means 'worldwide' in Italian and French) delivers multiple high-quality audio services using over existing bands, including those used for AM and FM. It can exploit the huge reach of MW and SW, covering thousands of square miles with robust, high-quality signals, using a single transmitter. The DRM consortium launched several new initiatives at the 2023 IBC *(International Broadcasters Convention)* in Amsterdam in September 2023 which were well received.

DRM test transmissions have been taking place for 25 years now but, so far, it's only in India where they have begun to replace regular services. The BBC World Service and some other international broadcasters use the method on some of their HF broadcasts. Reception reports and the received audio quality have been excellent but so far, there has not been a rush to buy receivers.

One of the reasons has been the sheer cost of these; the only ones that came to market were priced around €800, yet offered no new programmes. The general public do not buy a radio on which they can't hear anything different. 99% of the population are not interested in technical advancement or the audio fidelity of programmes – they are attracted only to new features and programmes.

ONLINE Radio

VHF Band III is a higher frequency band around 200MHz, where digital broadcasting was launched in the 1990s. Using digital modulation methods, multiple stations can be transmitted by one transmitter, making the platform potentially more economical. The aerials needed are only half the size of Band II antennas used at FM.

The VHF bands offer the ability to tightly focus transmissions into densely populated areas but there are also some disadvantages in digital transmissions at VHF:

1. the signals are more susceptible to degradation by topography; even trees can limit coverage.
2. Changing meteorological conditions (e.g., heavy rain) can affect coverage.
3. Signals travel much shorter distances and some DAB transmitters barely reach a few miles. Useful in reusing the frequency for more stations, but frustrating for listeners when signals disappear across a room.

When DAB was first launched it simply carried existing radio stations, and the receivers were very expensive. It wasn't until new programmes were added that it had any traction at all, but it was still a field on which only the big boys could play – the BBC and the two largest commercial radio companies.

SMALL SCALE DAB

This new initiative was designed to open DAB to a new tier of small-scale broadcasters broadcasting local community-focused radio stations. OFCOM began experimenting with small-scale DAB in 2013 to test the technology and prove its efficacy. Pioneered by Ofcom engineer, Rash Mustapha, the experiment used some freely available software and computer technology to transmit digital radio services and broadcast to a relatively small geographic area. It allows stations to use inexpensive equipment to transmit for far less money than had previously been possible.

More details of the DAB and DAB+, in the book available at:
https://worldofradio.co.uk/DAB.html

ONLINE Radio

ONLINE RADIO is 'over the air' too!

UHF frequencies above 500 MHz have been used for TV for the last fifty years, as well as communications networks such as mobile phones and localised WiFi networks. In most homes, WiFi is transmitted at around 2000 and 5000 MHz (known as 2 and 5 GigaHertz). This is the medium used to get online radio to listeners, so online radio could strictly be described as being a user of 'through the air" radio frequencies too.

Over the last hundred years, radio transmission has always been a little confusing with many unusual nature phenomena.

Many radio stations are now relayed by TV multiplexes, both terrestrially and on satellite. *FreeSat, Sky,* and the regular DTT services in the UK all carry radio stations, but all are relays of existing stations and none are 'stand-alone' broadcasters using just those platforms, despite credible size audiences being reached via TV.

Chapter 4

LEGAL & Licensing

Broadcasting is a powerful vehicle for reaching the hearts and minds of others and for exerting immense influence on beliefs, habits and commerce. It is right therefore that broadcasting is subject to some rules and regulations, to protect society as a whole; only a fool would dispute that. Sadly, the law has been abused and unfairly used to suppress various parts of society and regulate tightly who is allowed to participate in communicating how, why and when, with others.

Even legal draughtsmen, the people who assemble the laws and licence regulations, are not infallible. They often fail to anticipate the thirst for licences among prospective broadcasters and omit to make the rules sufficiently watertight. Some legislation doesn't just leak however, you could drive quite a big boat through it, often literally. While the UK's marine Offences Act supposedly outlawed radio ships off the British coast, the law was weak and had several lacunas. These were wide enough to permit radio ships to continue operating for another quarter of a century with relative impunity. We claim credit for some of this, but in reality, it was the lawyers who made our work possible.

ONLINE Radio

Broadcast Licences
The rules for licensing Online Radio broadcasting vary from country to country. In many territories you do not need a broadcast licence to operate as an online broadcasting station, while in some countries you are not allowed to stream radio or video broadcasts at all. In the UK, only 'over the air' transmissions require licenses (as of late 2023) and broadcasts which are made online, i.e. over the internet, do not currently require a licence.

Broadcast licences have been withheld from private ownership in many countries for many decades, except the USA. In the UK a very limited form of private radio was licensed since the 1970s, but this has been rigidly controlled and only large well-founded companies were allowed to broadcast. A strict regime of non-competition was imposed by the *Independent Radio Authority*. It tightly regulated all radio and kept it from the hands of everyone except the powerful and the well-connected leading to a lack of growth. This continues to be the case, where the only licences for private radio stations are for small community and non-profit making stations. Internet Radio remains the only unrestricted outlet for broadcasters.

A similar pattern emerged in many other countries such as the Netherlands, whose government has only permitted privately owned radio since 1994. There too, the majority of radio stations are now run by a small number of large radio groups.

If you want to operate a traditional or terrestrial radio station in most countries you must obtain a broadcast licence from each sovereign country's broadcast authority. Details of licensing requirements can be obtained from an international broadcast consultant, such as Worldwide Broadcast Consultants, or direct from each regulator individually.

https://WorldofRadio.co.uk/WBC.html

ONLINE Radio

Content Licence
In addition to a 'transmission' licence, radio stations may need a content licence to transmit news and information. Most Western countries do not have such restrictions and allow any of their citizens to broadcast any verbal content, so long as it is within the general free speech guidelines of 'decent, legal, honest and truthful.'

A content licence is not required in the UK, but beware of the many laws that restrict legally what you may say or publish. In general these cover the 'common sense' rules of

LEGAL DECENT HONEST TRUTHFUL.

The laws of defamation and libel apply to broadcasts, just as they would to statements or claims made anywhere else in public, such as in the press.

Legal advice on journalism
Essential Radio Journalism is the title of comprehensive working manual for radio journalists as well as a textbook for broadcast journalism students. It contains practical advice for gathering, reporting, writing, editing and presenting, the news, alongside media law and ethics.

Written by radio guru **Paul Chantler**, the book has a wealth of 'inside' information, checklists and on-the-job advice that you can immediately put to use whether you are in your first job or have several years of experience. A book to inspire responsible, accurate and exceptional journalism skills.

Essential Radio Journalism
Paul Chantler & Peter Stuart

Available as a Kindle or a Paperback : *https://amzn.to/3QjzfHb*

ONLINE Radio

Copyright Licences

The laws of almost every country say that, if you create something, such as a painting, a music recording, or a book, then you OWN the intellectual rights in that work. The owner of any creative work, or the IP rights in it, has the right to decide how and where it's used. Anyone else using it, especially in business where they might benefit from it, such as music within a radio programme, must be licensed by the owner.

Individual agreements between IP owners and users would be a nightmare, so the rights owners are represented by societies of the IP owners. Licences are issued by them and the royalties collected for the use of works on behalf of the owners.

Music Copyright

Recorded music is used by a wide range of businesses to attract customers, drive spending and motivate employees. In most countries, a user must pay a royalty for the use of music in the course of business. Globally, copyright forms about 8% of all from music, around $950 million. That compares to around 46% coming from digital products and 46% from physical products (more detailed analysis in the *IPFI Global Music Report*, available on their website. IFPI are global and based in Switzerland.

Radio stations must pay for their use of music, just like bars, clubs, restaurants, gyms, etc. Music content is copyrighted and radio stations must obtain a licence to play commercially-issued music during broadcasts. What is "commercially issued? If you can buy a copy in a store, either physically or downloaded, then it is 'commercially issued'.

www.ifpi.org

Copyrights are often referred to as royalties, except by people in the music business who call them 'rights' and by lawyers who use the term 'I.P.' (Intellectual Property).

Commercially recorded music includes any items bought as a CD or a downloaded. There are two types of copyright in most music:

    A. the performance copyright granted by the performer and the company who made the recording, which is collected by PPL.
    B. the authors of the music and any lyric collected by the PRS.

ONLINE Radio

Copyright Territories

The world does not have totally international legislation; each sovereign nation sets its own laws. While many follow the English legal system, other countries set their own rules and tariffs. Some countries choose not to enforce music copyright at all. As Internet Radio streaming is global in reach, there is nothing to stop radio stations establishing their base in whatever country they wish, and make a choice, perhaps considering only how 'lax' copyright law and collection enforcement is.

If one copyright or royalty fee could be applied worldwide it would solve a lot of problems. The argument that 'one standard fee' would be unfair in poor countries only holds water where the fee paid for use of a copyright is a fixed amount. If however the copyright fee was set as a reasonable percentage of turnover, then the system would be fairer.

Unfortunately, the international copyright organisations have not yet grasped this, just as the music business 'missed the boat' and failed to see the point on cassettes, CDs and downloads. The quest for maximising profits appears to have blinded them. Rather than have a small and sustainable share of revenues and keep the 'golden goose' laying, the collectors seem to prefer their clients receive nothing.

In recent years the music companies insisted they would not licence the use of music downloads and this attitude crippled the start of music, often to the detriment of the very musicians they claimed to be protecting. Years of protracted negotiations between the big music companies, such as Universal and Sony, saw commercial sense prevail, eventually, and licences were issued which has resulted in services such as Spotify and practical rates for Online Radio stations to use commercial music in their programmes.

Payment of Royalties.

Online radio stations must be licensed to play music, live or recorded. The actual rate paid is decided by tribunals and agreed by bodies representing the users and collected by societies acting for the owners of the rights.

Content that has been released under creative commons or is regarded as being in the public domain, or that music has fallen out of copyright protection (due to its age) does not have any intellectual property (IP) protection.

ONLINE Radio

Royalty Collection

Copyrights are administered by different companies which differ in each territory or country. The main copyright bodies or collection agencies, and their various functions, in the UK are the PPL, PRS and MCPS.

(PPL) Phonographic Performance Ltd

Their licences authorise the use of recorded music that is readily available in the UK. A PPL licence is not required if the copyright in ALL the recorded music a station plays is not owned by any PPL members. a small number of stations do play only 'unsigned' artists and so are exempt. *Radio Six*, broadcasting from Scotland, is in this category. The basic fee structure below applies to all stations, except 'online' which PPL calls 'linear webcasting' (see overleaf).

PPL's Radio Licences

PPL issue several different licences for radio stations which also authorise stations to stream their service to many other countries. The PPL licence does not authorise a radio station to play music in public places such as a bar or restaurant.

Standard Licences for non-commercial stations.
Two licences are issued to stations classified as non-commercial which depend on how many performances (plays of recorded music) are made.

A **limited** 'non-commercial' licence applies for less than 270,000 'performances' a year with revenues under £5,000. The Fee is £205

A **basic** 'non-commercial' licence is for stations with revenues under £5,000 a year but they may make unlimited performances. Stations must make a quarterly return and confirm listener hours and the average number of tracks played per hour. The Fee is £400.

Standard Licence (Commercial)

The Standard 'commercial' licence covers Online Radio stations with revenues in excess of £5,000 a year. The reporting requirements are the same as for NC stations (above) and the Fee is £690 pa.

All the fees are subject to VAT and are recoupable against a rate-per-performance of £0.000757, plus an administration fee of £100 per year for PPL applying the IFPI webcasting reciprocal and licensing online radio streams into multiple territories.

ONLINE Radio

PPL's 'Linear Webcast' licence

Stations including recorded music in an online radio station or online service broadcasting from the UK, need a PPL Linear Webcast licence. This licences services (for example, an online radio station or simulcast) where the audience cannot skip tracks or pausing the broadcast.

The Linear Webcast Licence is subject to an application process, and any licences issued are subject to a set of terms and conditions. Stations are graded as to the number of performances per year:

PPL BAND	Annual performance per channel.	FEE
BAND A	< 150,000 performances	£177
BAND B	150,001 to 270,000 performances	£295
BAND C	270,001 upwards performances UK and foreign audiences Generate over £5,000 in revenue	£591

The majority of online radio stations are covered by a Band C licence with only the very smallest operations being able to claim they are Band A or Band B category. A station playing an average of 15 tracks an hour will soon accumulate 131,400 performances a year with only one listener on overage – at just two listeners on average it is already at the 270,000 limit.

Radio stations need to report their use of music either quarterly, or twice a year in a Broadcast Data Report, comprising two figures:
- The average number of tracks webcast per hour
- The total listener hours in that period.

These are standard statistics that are typically captured by a streaming provider. The listener hours may be called something slightly different such as 'Aggregate Tuning Hours' or 'ATH', or simply 'TLH'.
See the following guidance for more details:
https://www.ppluk.com/music-licensing/broadcast-data-reporting/
PPL licence applications can be made at:
https://broadcasting.ppluk.com/buyalicence/s/

ONLINE Radio

PRS. Performing Rights Society

The performance word is a little misleading as this licence, used by the PRS and MCPS jointly, is permission from the composer of music and the writer of any lyric content. PRS payments are collected on behalf of songwriters, composers and publishers. Payments to the PRS are due whenever music is played in a broadcast, radio or television, is streamed or downloaded or is performed in public, such as by a covers band in a venue or at a live concert.

Playing recorded music without the appropriate copyright licences is an infringement of copyright and the copyright bodies can take legal action to recover lost fees. A court can order the broadcaster to pay licence fees for previous periods of broadcasting as well as PPL's legal costs. The court can also issue an injunction forcing the broadcaster to cease playing recorded music until this is done. Ignoring any court order can be held to be contempt of court, which could lead to a broadcaster being imprisoned.

Mechanical Copyright Protection Society

The MCPS is an organisation authorised to collect royalties on behalf of music publishers, songwriters and composers, when their music is reproduced, in any format, including online radio.
The main three areas are when music is copied as a physical product, such as a CD or DVD, when it is streamed or downloaded or when used in a film, TV or during a radio programme.

PRS MCPS Licences

The Performing Rights Society and the Mechanical Copyright Protection Services combined their administration teams a few years ago to simplify the procedure for issuing music royalty licences. The societies now issue combined licences for radio stations to pay royalties to the writers and composers of music.

https://www.prsformusic.com

Smaller online music services whose revenue is less than £12,000 per year need a **Limited Online Music Licence** which covers on-demand streaming, permanent downloads, podcasts, webcasting, karaoke services and general entertainment. You can download the LOML guide from the *PRSforMusic* website. Fees start at £130 per month for online radio stations. More details from the PRS (020 3741 3888) or by emailing them:

applications@prsformusic.com

ONLINE Radio

Limited Online Music Licence
The LOML is for small digital services in the UK that offer on-demand streaming, downloads, podcasts, webcasting and general entertainment to an audience in the UK. Services are classed as being 'small' if the gross revenues earned are less than £12,500. The fees are based on how much music is used, and start at £154 a year. It's an annual blanket licence, but the licensee must complete a report form at the end of the period confirming how much music is used. PRS staff advice: 020 3741 3888

Administration Fees
In addition to the advance fee and any streaming royalties that may be due, licensed services must also pay to PPL an administration fee of £100 per year to cover PPL's accounting of royalties to the music licensing companies in each of the territories covered by the licence. Where a service is restricted to UK listeners only, by technical measures such as 'geo-locking', the administration fee is waived. For more details, see the PPL web site.

The LOML licences do not cover
- Sound Recording rights. (see *ppluk.com* for details)
- Services operating outside the UK
- Synchronisation rights
- Use of Music in commercials sponsorship or any advertising.

Overseas Coverage
PPL has entered into a reciprocal arrangement with overseas music licensing companies that represent recording rights holders such as record labels and performers. Many major markets are covered, though not the USA, nor France. The administrative fee for broadcasting to territories outside the UK under the IFPI Webcasting reciprocal arrangements starts at £68 per annum for a small, single-channel, online radio (linear webcast) station.

The multi-territory licences are handled by a separate department and fuller details can be found on the territories page of PPL website.

https://bit.ly/3s5Foxy

ONLINE Radio

PROGRAMME LICENCES (OFCOM)

In order to be carried on any of the radio-delivered channels (such as FM, or DAB) regulated by OFCOM, a radio station must obtain a **Digital Sound Programme Licence.** These are available "on demand", usually within four weeks and full details are available on the OFCOM website. Radio stations can apply for a programme licence at any time and operators need not obtain transmission facilities before a licence can be awarded.

There are two types of licences available:
DSP (Digital Sound Programme) for regular stations, and the
CDSP (Community DSP) for *not-for-profit* operators.

OFCOM charges an Application Fee of £250 for a DSP or a CDSP programme licence, then an annual fee of £100 is charged for its renewal. Note that this DSP licence does not confer any rights to broadcast nor any rights to carriage on any transmitter or multiplex.

A digital sound programme (DSP) licence is required by anyone who wishes to broadcast a sound programme service on a digital multiplex (DAB or SS-DAB) whether this service is unique to digital or a simultaneous broadcast of an existing analogue, satellite or cable radio service.

One digital sound programme (DSP) licence covers all the services provided by the licensee on any number of multiplexes or webcasts (online radio) but separate licences are required for local and national stations broadcast on radio transmitters. A national DSP licence is also required if the service is to be broadcast on DTT Freeview.

More details about OFCOM's DSP licence can be found on their web site, along with the Application Form, which they prefer to be submitted by email. Full details are in a briefing paper that can be obtained at this URL:
https://bit.ly/3FpTzAy

Over-regulation often stifles creative types and certainly hampers the growth of new media. In 2020, Steve Mason, launched *Radio Phoenix International* which proclaims it is proud to be a free radio station. He explains: "Online Radio is cheaper than the regular airwaves which are too tightly controlled by OFCOM. As a medium Online is more fun, because it's not regulated into the floor so can play a much wider range of music than the established stations and their 'Top 5' only playlists."

ONLINE Radio

Total Listening Hours

TLH stands for Total Listening Hours, a very important number, as not only is it the basis that radio stations use for charging for commercial air time, it is also the basis for the charges for calculating the costs of carriage of the signal by hosting companies. The TLH figure also forms the basis for payment of music royalties. TLH is a count of the amount of time spent listening to a radio station, per listener. It is calculated by the number of connections to a stream multiplied by the number of hours they are connected for. Ten listeners each listening for ten hours equals a TLH of 100 hours.

The TLH amount is reported as a statistic in the control panels of most of the Internet Radio systems. It is a vital tool for measuring listening totals which is reported to the music copyright collection agencies. They calculate the amount of royalties that stations must pay over to them.

Inter-active Streaming, or not?

Streaming falls into two distinct camps for copyright charging purposes; either Interactive or Non-interactive, depending on whether each listener is involved in selection of the music. A Non-interactive service is a simple broadcast where the music is chosen, perhaps by a DJ or an automation system. The reasoning is that if people can simply dial up a service to play the music they wanted, sales of recorded music will be affected. Clearly there had to be a distinction between the two situations and different pricing structures prevail.

Rules were also implemented into non-interactive services to prevent any artist's material being repeated more than 4 times in a three-hour period, and never more than three tracks in a row by one artist. For any particular album, the limit of three tracks in as many hours, or two consecutive tracks, is imposed.

This reason is why many stations were unable to play a full day, or even an hour, of David Bowie tracks when he died in January 2016. Unless specific waivers are put in place between an artiste and each station using it, lengthy tributes are against the rules. David Bowie was always a firm advocate of copyright being abolished and was the first to release material on the internet only, in 1996. See this web page about it.

https://worldofradio.co.uk/Bowie.html

ONLINE Radio

ONLINE Radio

Chapter 5

Studio Equipment

Less than a generation ago, radio and TV studios would use very expensive equipment; a typical radio studio, even those used for local or community radio, would cost between £20,000 and £150,000 to build. This can now be achieved for just a few thousand pounds and the resulting studio will provide much better audio quality and be much easier to operate than the complex studio setups used previously.

Digitisation has revolutionised radio studios and programme origination. Until very recently all sound sources in radio studios were analogue, which means they sent a constant stream of varying audio information to the user. Sound sources were traditionally quite large and they used separate cables for each source, until the point of mixing. Digital has decimated the size and cost of almost every item.

Digital audio has seen some profound effects in four distinct areas of the operation of radio stations. These are felt in control, equipment miniaturisation and especially in cost.

Audio production Many sources have changed beyond all recognition, especially in size. The adaption of digital techniques, especially the use of compression, renders files much smaller and the ability to record on very small (physically) media.

ONLINE Radio

Control of equipment has become easier, with remote operation of sources being simple to achieve using digital signals, sequences of 1s and 1s.

Administration and record keeping. Vital in tracking royalty liabilities.

Transmission: the medium of getting the finished programme (audio and video) to listeners. In Online Radio, the system is, at the moment, one of **unicasting** the programme from servers to the playback equipment and this would not be possible if the material were in analogue form. Digitisation, bringing the possibility of compression and ensuing reduction in file sizes, makes it all possible.

The ways in which radio studios operate and the methods by which radio programmes of all types are produced has undergone remarkable transformations. Methods of access and playback have changed, although generally the manufacturers have tried to keep the appearance and methods the same, to encourage the changeover from analogue to digital.

The playout equipment has becoming far more efficient, smaller and thus very much cheaper. They are also much more easily controlled and are smaller, faster and cheaper too. Instead of having to be adept at equipment operation or the technical parameters a radio presenter is free to focus on the programme's content.

Even something as simple as the cabling between various items of equipment and studios is improved and changed. Previously all interconnection of audio, and often of switching circuitry too, had to be in screened audio quality cable, with individual connectors terminating each cable run. Dozens of cables have now been shared by a very simple network of Ethernet cables, sometimes referred to as "Cat 5" (a category of cable).

Cabling a studio previously took days or even a few weeks to complete, but any item or complete studio can now be connected into a studio or a complete broadcast centre by one simple Ethernet cable. (An Ethernet cable is simply four pairs of twisted cable in one encapsulation, with a small modular 'RJ' connector at each end. It can carry multiple signals, but take care to select the correct one as there are some variations and it's important to have the right cable correctly terminated in the right connector).

Traditionally, the 'sources' of programme material came from hardware. In the first radio stations, the original equipment was simply a few microphones. Where a programme called for music performances, even of

ONLINE Radio

short length, it would be performed live in the studio. To play recorded music a microphone would be held in the horn of a gramophone which was an acoustic device, with a needle crudely vibrating a diaphragm. The sound created from those vibrations and transferred into the air acoustically, was converted into electrical signals by the microphone, just as live voices were.

Radio studios developed gramophone pickups to give electrical signals, and then wire recorders were introduced; they were the granddaddy of the now increasingly obsolete tape recorder. The item that's been around for over a century has been the master hub or switching centre, which evolved into the mixing console for controlling and merging audio signals.

Studios today are a world away from just a decade or two ago. First the playback equipment (grams, tapes, carts, etc) were replaced by migrating them onto hard drives and then onto solid-state drives. These could be securely stored in the station's racks room. The programme host's desk simply comprises a few monitor screens, a mixer and a microphone.

In the next stage, presenters were packed off to voice their input remotely, often from home, in a process called *voice tracking*. Their voice could be inserted by a computer running some automation software. This has become common-place in recent years with the number of staff needed at the radio station's hub, the studio centre, dwindling.

Today's latest radio stations don't need a racks room, any equipment or even a studio as all the playout kit is cloud-based and controlled from the station operator's computer. A PC, laptop or a mobile phone can remotely command mixing, levels, EQ and switching; all jobs which were previously done on a mixing console and other kit located at the studio centre.

These latest automatic systems, such as British company Broadcast Radio Ltd's new ***Myriad Cloud Radio,*** can be leased for just a few pounds a day. Stations can get a generous discount by mentioning this book; see p174 for the voucher code.

Studio with only monitor screens, mic and mixer
Ray Robinson

ONLINE Radio

Assembling a radio studio

At its most basic, a radio station could be simply broadcast a play out of continuous music, needing little more than a simple music player. A small iPod would suffice and these have been used these in times of emergency. Most radio stations however will also need announcements, so that means having at least one microphone. The station then needs suitable control equipment to switch and mix the microphone levels with other equipment, such as music players or other pre-recorded playout equipment.

Radio programming demands proper production, so several items of both playout and recording equipment are necessary to make smooth transitions between items. Microphones simply change acoustic audio signals into an electronic form and its usually necessary to apply some level of processing to the signals, to colour the sound to be comfortable for listening. This might be limiting or compression, equalisation, or some gating to remove noise.

All the above equipment can be bought almost anywhere, but the really good quality and 'controllable' equipment needed by professionals usually costs a bit more than that found in most homes, often referred to as domestic equipment. Buying the studio equipment and other items of hardware might be best with some input from a reputable broadcast consultant, see Chapter 12 for details of WBC and others.

A god quality mixer with flexibility needed to produce radio programmes usually costs a couple of thousand pounds. Recently, RØDE and others have produced "all in one" mixers for about £700 that skilled hands can produce broadcast quality programmes, even live to air. The Rødecaster Pro is an easily mastered model and there's a full web page about it here:
https://worldofradio.co.uk/RODECasterPro.html

Digital Mixing

Most radio stations today are migrating to digital mixing, where all the audio is converted to digital signals making it much easier to interconnect. All the audio in a radio station can be run on just one Ethernet network, reducing the amount of cable and connections to a small fraction. Fewer connections will usually mean that there is less to go wrong; one of the biggest faults in audio studios would be from poor connections. While good ethernet connections remain important, the likelihood of scratchy audio from poorly fitting cable connections is much reduced.

ONLINE Radio

Telephone Equipment

If you plan having any participation using phone lines, a call handler will be useful. This was previously known as TBU. This is a telephone balancing unit, on which you could (eventually!) get a reasonable balance between the caller and the calling parties. Modern phone units not only balance the line up to give you 'good audio' on the air and suppress echoes and other bad sound effects, but also include a recorder section. With this you can pre-record calls and carry out any editing before putting it onto the air, usually via the library.

These previously had to be done with open reel tape which often resulted in yards of tape all over the floor, a couple of cut fingers and invariably took forever. To perform even the simplest of edits. Now, with solid state recording, it's all accomplished much quicker, easier, with more accuracy and a lot more reliably too!

ipDTL

Today it's possible to use several web browsers to bring in audio over the internet. British company **In:Quality** has developed and operates a worldwide network for the real time transmission of professional audio.
ipDTL is a system which allows radio and podcast presenters to connect with studios, co-presenters, producers and guests. It requires only a computer with an internet connection, and a USB mic or audio interface.

It's organised through a British company called **In:Quality Limited** which has developed and operates a worldwide network for the real time transmission of professional audio.

Broadcast quality audio using Opus & SIP

In its simplest form, ipDTL connects one computer to another, giving 2-way audio between web browsers, using the wideband low delay Opus codec. SIP is the protocol for connecting between studios, which has replaced ISDN. ipDTL includes SIP calling as standard. It can also make phone and ISDN calls.

ipDTL users pay a monthly subscription from as low as £10 a month but radio stations can take a three days trial for just £1, to be certain that it works for them.

ONLINE Radio

ipDTL ~ Internet Protocol Down The Line

Established in 2013 in the North of England, its founder was previously a-presenter and an engineer at the BBC, Kevin Leach. The system is MacOS, Windows, Linux and Chrome compatible; all you need is a decent broadband connection. The system is rented on a monthly basis, although you can also buy a 'day pass'

Kevin Leach, founder

In Quality Ltd

In Quality Ltd also offer a **hybrIP** phone-In system which can replace the traditional studio TBU (Telephone Balance Unit) with ipDTL integration and call screening. It can be used for
- Talkshow Management
- Call Screening
- Live Call Handling Hybrid Replacement

This is the system used at *Virgin Radio* every weekday for the Chris Evans morning show. Details are here: *https://hybrip.com*

VOIP systems offer a way of doing so, and IP-DTL is worth researching too. In its simplest form, ipDTL connects one computer with another giving you a two-way audio link using a wide-band, low delay *Opus* codec.

CODEC IP-DTL can be used for live interviews, OBs, voice sessions and options such as multitrack WAV recording. You can bring in up to six remote sources at once on a single browser, so the possibilities and versatility are immense.

ONLINE Radio

Audio Storage

The most noticeable effect of the changes brought about by digital has been the methods of programme item storage, often called Asset Management. Previously HUGE libraries, that what once took up several huge rooms of mechanical racking, can now be easily accommodated on a couple of magnetic hard drives. These offer up to 1 Gigabyte of storage, and several are usually connected together in an array, offering several gigs of storage, more than sufficient to store over 100,000 music tracks.

SSD

Solid State Drive storage is the latest way of storing audio library material, or items for broadcast, or other sort of data. While magnetic hard drives are still the most cost efficient, especially when considering volume, it is possible to use something better and faster.

Totally solid-state storage (often called 'flash drives') has the audio, in the form of digital data, burned onto a network of circuits contained within silicon microchips. They can be as convenient and small as those built onto a USB connector and known as 'thumb drives', or can be a mass of chips many gigabytes or even terabytes in size. Usually for library purpose they are mounted as an array of several SSDs with switching to ensure that data is spread or even duplicated over several drives, avoiding the anguish when drives fail. They can do, eventually, but SSDs are far more reliable than the old Hard Drives, which are a stack of spinning magnetic discs.

Using of these Solid-State Storage Drives removes the risk of mechanical breakdown, which is currently the biggest cause of failure in data storage.

The amount of storage space required is set to plummet still further as new methods come on stream. The huge increases in size and reliability in hard drives mean that complete music collections can easily be stored on a single unit, and at far better quality than MP3, the usual standard that's tolerated by consumers. Many radio stations keep all audio as uncompressed audio; usually in the AIFF or WAV formats.

The cost of SSD and the older magnetic hard drives is in a constant state of flux. We monitor these and the latest bargains and details are on our special web page.

https://worldofradio.co.uk/drives.html

ONLINE Radio

It's appropriate to be aware of and perhaps consider as archives, some of the following items of audio storage:

TAPE

CC

The first tape recorders were huge spools of metal wire onto which the signal was recorded by magnetism. It was so noisy that it had to be housed in a separate, deadened room. These were much improved by baking small particles of oxide onto paper and later onto plastic tape, usually Mylar, a brand of PET from DUPONT. Refinement of the 'tape' and recording heads led to improvement in audio quality and a reduction in size and speed.

In the early 1960s **Compact Cassettes** were developed by Philips. These were a pre-packed tape in a small plastic cartridge. They became popular in domestic use, but the very slow speed and lack of accurate control rendered them unsuitable for broadcast use.

The Fidelipac tape cartridge was introduced in 1959 by Collins (at one time they a well-known US manufacturer of most items of radio equipment). Fidelipac, or **NAB Carts**, ran at the more 'studio standard' speed of 7½ ips and each 4-inch cartridge could hold enough specially lubricated tape to run up for up to 10 minutes. It was ideal for playing commercials, items of music and ID jingles.

ONLINE Radio

Carts featured three cue tones to make it easier to use in broadcast studios. They stopped and started the tape motors where necessary and, in some cases, had a third (tertiary) tone for fast forward at the end of a piece. Some of the tones needed very careful filtering to exclude them from the audio path. With some old analogue automation systems you could hear the tones (especially the secondary tone at 150 Hz and the tertiary (fast forward) tone at 8KHz) along with the programme's audio.

NAB Cart players are often erroneously referred to as a *Spotmaster*, which is a trade name of one of the first company manufacturers. Others were made by Tapecaster, Gates, ADC and ITC. In the UK, similar machines were made by Plessey and Sonifex.

Radio studios once were home to several *Lazy Susans*, filled with carts, especially well-funded stations where the music was played in from them, eliminating record wear and jumps. All now been replaced by racks of drives, which are usually unseen but accessed via monitors.

In the early 1960s the record companies began offering many of their albums for commercial retail in a slower running package called an **8 Track Cartridge**. This was a simpler version that ran at half speed enabling a full album to crammed in over the 8 tracks. Within twenty years it had been overtaken by the compact cassette and then the CD, but for a while was popular in the USA, mainly for in-car use.

**'Open Reel' equipment is still revered today
as it gives the most faithful and purest form of audio.**

DAT

Digital Audio Tape cartridges were very small yet could offer several hours continuous recording. The tiny DAT cartridges are the smallest form of encapsulation yet used in domestic or professional audio, measuring just 73 mm × 54 mm. It became popular in professional recording studios for a time as it could offer better quality than the Compact Disc. Sony's MINIDISC system was popular in Japan but it never really took off in professional audio use, especially once hard drives increased in storage time and reliability.

ONLINE Radio

MUSIC LIBRARIES

Networked audio storage systems can enable multiple operators to simultaneously access and used audio libraries and these have now become *de rigueur* in radio studios. These were developed in the 1990s using arrays of hard drives (HDs) but since the cost of solid-state drives (SSDs) fell in the 2st century, these have now become the library of choice, capable of holding thousands of hours of material. **Networked Attached Storage Drives** (NAS) usually have multiple drives and make several copies, giving redundancy and mitigating for any disasters when drives fail.

GRAMOPHONE DISCS

Throughout the last century the gramophone record continued to be the main source of music recordings, with the only developments being the slower speeds and better quality that resulted from a change to microgroove in the 1950s, enabled long-playing albums. The 78 rpm discs which predated radio were obsolete by 1960 when the 45 become the standard for singles and 33 for albums.

Vinyl, an oil-based plastic, has had a revival over the last decade with high quality pressings made using thick vinyl for favourite albums. Many rare recordings are only available on old vinyl singles and albums from the 1960s and 70s and can command high prices at auction.

CD

In the early 1980s the Compact Disc was introduced. This was an optical device capable of storing up to 80 minutes of digital audio. Their small size meant they began replacing the Compact Cassette for commercial releases and then the vinyl disc. Later they were adapted to record data too; compressed digital music to the MP3 standard led to the discs being able to store 80 minutes of raw audio, or a massive 4 or 5 Gigabits of compressed audio, several days' worth! Being digital the discs could also be used for video as well as audio which meant a couple of hours of really good quality video.

Record collectors, disc jockeys and radio stations had already replaced their collections several times over a couple of generations. Ignoring the 'limited appeal' formats such the of DAT, Minidisc, etc, the cycle for domestic audio has been:

Gramophone → 8Track → Cassette → Compact Disc

ONLINE Radio

MICROPHONES

Amazon

The main item in any broadcast chain will be the microphone – even in a station that plays 100% music, it's likely that this will originally have been captured by a microphone originally, unless the station is a 100% totally synthesised electronic music operation.

There are many sources of microphones; the best advice is to get the best quality microphone affordable. Microphones now vary in cost from literally pennies, for the devices used in children's toys, to several thousand pounds for the top-quality microphones used in recording studios.

Getting the right microphone, i.e. the most suitable, for the job is important. Advice from a broadcast consultant with audio knowledge is the best route, however most suppliers offer different types. The *World of Radio* microphones web pages have many bargains suitable for use in radio studios plus links to over 40 different web sites offering advice.

http://worldofradio.co.uk/Microphones.html

The main manufacturers of microphone used in radio broadcast studios are AKG, Audio Technics, Behringer, Beyer Dynamic, Røde, Electro-Voice, Heil Sound, K & M and Neumann.

Simply using any old microphone, perhaps a simple one designed for a domestic tape recorder or worse, communications equipment such as a CB radio, will probably NOT give you the quality that your vocals deserve. To understand how microphones work it's easier to know a little about audio and acoustics.

Microphones capture acoustic sound waves by a diaphragm vibrating in sympathy with air movement. They are an 'acoustic to electric' transducer. There are many different types, both in construction and in purpose. We shall not look at the more specialised ones designed for particular music instruments, such as drum mikes, and so on, but concentrate on those used in radio studios.

The human voice is a complex mix of a range of frequencies, with the fundamental frequencies of adult male voice being from 85 to 180 Hertz (1 Hertz is a cycle per second) and women usually at the top end of this range, about 165 to 255 Hertz. The harmonics added to the fundamentals in modulation (speech, or other vocal sounds) are what adds the colour and texture to the sound. When harmonics are removed or altered in a processor, it can alter the sound of the voice considerably.

The human hearing range is traditionally regarded as being from 20 Hz up about 20 KHz, but most of these notes are outside the range of human voice production and are produced mechanically. Few musical instruments have a higher fundamental capability beyond a few kHz and the higher notes, above say 5kHz, (pipe organs and violins mainly) only produce the really high frequencies as harmonics of the fundamentals.

The human voice does the same; the fundamental notes made by the voicebox are coloured by altering the resonant cavities of the mouth to double, treble (and even more) the fundamental frequencies, the resultant complex sounds produced are rich in 'harmonics.

ONLINE Radio

The earliest microphones were the carbon granule type, effectively a container of small carbon granules with a couple of electrodes. A diaphragm responding to changes in air waves striking it changes the resistance of the granules. This varies their electrical resistance and so 'modulates' the electrical signal passing through the electrodes. Carbon microphones don't respond uniformly to all audible notes, but they are reasonable at the audio frequencies that make the human voice intelligible, those up to about 2.5kHz.

Initially the most common use of microphones was in telephone handsets. The mass-produced carbon inserts were unreliable, and the solution was to simply bang the handset firmly, which didn't do it much good, so they can to be made more and more robust! Long cables limit the range of frequencies that can be handled with the highest usable frequencies being kept to those below about 2.7kHz.

The result is what is usually called 'telephone audio' and it includes the frequencies that make human speech intelligible, so was fine for simple communications use. It sounds even worse as it does not carry the harmonic frequencies, which add depth, colour and tone to a voice.

Carbon granule microphones will not be found today in broadcasting. The quality is too poor and they have become regarded as legacy artefacts of a bygone age.

As early microphones did not produce the higher frequencies, little attempt was made to make other equipment in an audio chain responsive. Audio circuits actually suppressed the higher frequencies. Most audio equipment older than a few decades will probably be of poor quality and most likely be monaural.

ONLINE Radio

Microphone Developments

By the 1920s, radio broadcasting was demanding better audio quality and the moving coil microphone (like a loudspeaker, in reverse) appeared. It was followed by the ribbon microphone and later by several condenser types.

Condenser microphones give much better audio quality as they are smaller and can respond to the higher frequencies, and respond more evenly across the audio spectrum. Even higher quality microphones are now becoming available thanks to new types of microphones with a diaphragm of fibre optic. These microphones respond to air waves striking them which varies light from a laser. Using laser along fibre optic cable gives even better quality. The signal can also be manipulated, or processed, more readily.

Special 'music circuits' were also developed enabling a wider range of frequencies to be carried. The developments in audio came about as a result of the demands of the broadcasting industry and began in the 1920s.

Ribbon Microphones

These have a very thin ribbon of a conductive material suspended between two magnets. The air vibrates the ribbon and this modulated the current in it. Their main property is that they give very high quality, but are not seen often in radio studios as they have a bi-directional pattern. i.e. they pick up sound equally well from behind and the sides as they do in front.

They were also once called velocity mikes, as they responded linearly to the velocity of the sound wave. They were the main microphone used for symphony concerts and in broadcasting for many years, being found in the iconic BBC Type A and RCA microphones, now seeing a modest resurgence in use.

BBC

Condenser microphones

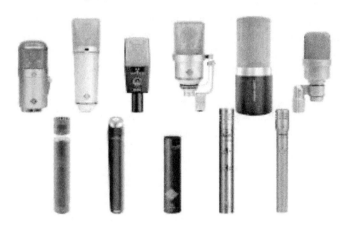

Amazon

Condenser microphones often have a bias charge applied, usually 48v DC. They are the most common in recording studios and radio work today. Another variation of these is the RF biassing type, which uses a small RF oscillator to bias the device, giving very sensitive microphones for really high-quality work.

They work by reacting to acoustic waves hitting their diaphragm. Vibrations on this cause the distance between the microphone's two plates to change which varies the capacitance to fluctuate in sympathy.

They are sometimes a little more fragile than other types of microphones and can be affected by big changes in humidity and also by temperature. Very cheap condenser microphones often use low quality components in the circuitry they contain and this can manifest as noise on the audio.

There are many condenser microphones on the market, including the most expensive such as those from Neumann that can cost several thousands of pounds. Good quality condenser mics can however be had for around £100 upwards. See here for a selection on Amazon: *https://amzn.to/493YZ1w*

Amazon

ONLINE Radio

Dynamic Microphones are the cheapest to be found and are usually more robust than other types. They are basically a small diaphragm surrounded by a coil. Airwaves striking the diaphragm cause the coil to move back and forth within a magnetic field and the minute currents generated are have the sound modulated on them. Dynamic microphones are suitable for broadcast use, especially when by energetic DJs who may be speaking with a music background.

Shure

The **Shure SM58** is probably one of the most familiar microphones, seen on stages and television in the firm grip of performers, or being abused and swung around their heads! You won't be surprised to learn that there is an active market in replacement parts for these mics, the microphone capsule of course, but also the iconic spherical metal mesh grille with a built-in pop filter shield can all be easily replaced.

Shure's SM58-LCE is perfect for radio, giving high-quality vocal reproduction, thanks to its tailored frequency response and brightened midrange which is ideal for punchy radio voice announcing. This model has a uniform cardioid polar pattern and it delivers a warm, crisp sound. The SM58 is a good solid mic' and costs around £100 here, complete with a leather pouch and bracket, a 3-pin XLR connector and the A25D swivel stand adaptor. *https://amzn.to/493YZ1w*

With a history of innovation going back to 1925, they are one of the best-known microphones and have lasted almost a hundred years thanks to their robust manufacture and sound quality.

As well as being more tolerant to physical abuse, dynamic mics can also handle sudden loud noises that condenser mics are a bit averse to. They will also have a better tolerance to signal overload. Dynamic microphones generally have a better tolerance to 'proximity effect' which makes the audio sound 'boomy' if the speaker is too close. Some mics, such as the **Electro-Voice RE20**, have a special 'variable D' design which all but eliminates proximity effect booming.

ONLINE Radio

Response

There are two main response properties the microphone purchaser should be concerned with. These are the **audio response** and the **pattern**.

The audio response is the band of frequencies which the microphone will respond to. The value is usually given a lower and upper frequency, often as low as 20Hz and as high as 20kHz, between which the microphone's ability to convert audio signals into electrical signals is roughly level.

The **pattern** is the directional sensitivity (or the 'directivity') of the microphone. It is obtained by measuring how well the microphone picks up signals from each direction around it and drawing these on a chart. The pattern of the levels of a particular level can be omni or bi-directional, it may be cardioid or some other intricate shape. If recording a large group of voices, it may be better to have an omni-directional pattern response, but if the sound is wanted from just one direction, then a unidirectional response is preferred.

Some microphones are often described as unidirectional, meaning they respond better to sound coming from just one direction. These have a cardioid pickup pattern, minimise background noise pick-up.

The property preferred for radio studio microphones is a cardioid response pattern, which is a technical description of the 'heart shaped' response area around it. The cardioid pattern can be thought of as simply rejecting noises that arrive 'off-axis' (the axis being a line through the centre of the microphone). These are much better for a self-drive operator who might be moving items and operating controls underneath it while talking, and probably shuffling papers.

Microphone patterns:

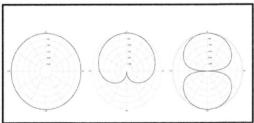

Omnidirectional.   Cardiod.   Bi-Directional

ONLINE Radio

Lavalier microphones

These are often known as a body mic', or a lapel mic' as that's where they are usually worn, enabling a presenter to do so many other things with their hands while talking. They are usually a very small condenser mic, or occasionally a ribbon mic', with a small clip for attaching to the presenter's clothing.

Lavalier mics are very useful when shooting video or interviewing, where a stick or hand-held mic 'may be distracting. They are not very expensive and are available from several different suppliers. There are many Lavalier mics available on Amazon. *https://amzn.to/3FrNqnd*

Amazon

Some Lavalier mics may have a small adjustable grill at the front. This varies the size of the cavity before the diaphragm and gives an amount of control over the audio quality. They can give a treble boost to the frequencies which tend to get muffled by clothing. If this is controlled by equalisation it could lead to more noise on the audio.

The Lavalier gets its name from its origins. The early ones were worn on a halyard around the neck, like the Lavalier jewellery or neck tie. The cardioid polar pattern usually found on a Lavalier reduces pickup of sounds from the sides and rear of the microphone, improving isolation of desired sound source.

Several types of suitable Lavalier microphones can be found here, including the appropriate accessories, including wireless links for them. To make a Lavalier mic REALLY useful however, and make it a proper high-end studio, we recommend using it with suitable software and a smartphone or a tablet. There are details below of a suitable app (*iRig Recorder*) that can do just this and a great offer on a pair of matched accompanying Lavalier microphones.

ONLINE Radio

Field Recording

Recording "in the field' or away from the studio about ten years needed a large UHER or similar 'transportable' recorder and probably several different audio processors. The equipment would have been heavy and usually needed a sound recording engineer to operate it all. Now all that's needed can be accommodated on an iPhone or iPad, or a dedicated hand-held recording device.

IK Multimedia are a well-known developer of all kinds of transducer and its likely your favourite band have some of their kit on stage or in the studio. They have developed an app called **iRig Recorder** that can be used to record broadcast quality audio, which it handles as professional quality .wav files.

The **iRig Recorder** app costs just £15 or so and it can also edit on the screen of an iPhone. There is also a FREE option which has less features but the audio quality is just as good; it's full 44 KHz 16-bit CD quality audio in stereo. The iRig app can be downloaded immediately from the Apple Store using this link: *https://apple.co/3QZCRyF*

The **iRig Mic Cast** is also very useful for interviews in the field, enabling you to monitor your audio as you record. The iRig Recorder app not only offers the possibility to cut, copy, loop etc a recoding, but has a suite of audio processing on board. They can add emphasis and process the audio quality to get the desired effect, all before sending the audio back to base either using a phone's internet connection, or the audio can be transferred direct over WiFi.
https://amzn.to/3QOGKWN

IK Multimedia have now developed an **iRig LavMic** which is just coming on the market. Its chainable and has built-in monitoring, making it an ideal hands-free microphone for just £45 suitable for mobile audio broadcasting and recording. For doing interviews 'in the field' the *iRig LavMic* is ideal and is available in a twin pack; a pair of units for £85. Now being used by many TV reporters in interviews.

https://amzn.to/46Ylcwc

ONLINE Radio

Radio studio microphones

Røde are an Australian company famous for their excellent microphones. Three of their special broadcast microphones are of interest; the *Broadcaster*, the *Procaster* and the *Podcaster*. All are metal cased, so very robust.

The Broadcaster has a wide frequency response and low levels of distortion. It also has a unique built in "on the air" light and an internal pop filter and a switch-able 'high pass' filter to remove low frequency rumble. The *Broadcaster* microphone may well outlast most radio stations and comes with a 10-year guarantee. It's about a tenth the cost of the industry's "top quality" Neuman U87 mic.　　　　　　　　*https://amzn.to/3tEjotV*

Røde's **Podcaster** microphones have a USB connection, which can be plugged straight into a Mac or a PC. They have their own audiophile quality 18-bit resolution, 48kHz sampling A/D converter which processes your mic audio inside the unit, by passing the computer's sound controller. Usually £160, there's 50% discount here.　　　　　　*https://amzn.to/40e3DpG*

Røde's **Procaster also** offers broadcast sound quality thanks to a large diaphragm. It's optimised for voice frequencies, is designed for close-up use and has a solid metal casing.　　　　　　　　*https://amzn.to/3tToq65*

The **Røde** company was set up by two emigrants from Sweden, Astrid and Henry Freedman, whose expertise in PA systems is legendary. Røde make many accessories including wireless remote transmitters and receivers. Røde have received lots of awards and are found in stations worldwide.

Amazon

ONLINE Radio

SHURE

A manufacturer known to most vocalists and broadcasters, their SM7 is designed for broadcast and has a large diaphragm as well as a flat response, so all frequencies are picked up equally. This model also has switches on the mic to roll off the bass response for speakers with very deep voices, as well as a control for mid-range enhancement which gives a useful boost to the frequencies which add most intelligibility.

Shure's SM57 is a relatively cheap model, costing just £109 here. While they are often found on stages or in music studios to mic up drumkits, etc, they can be very effective for radio studio use too. The SM58 is the familiar 'ball head' microphone and like all Shure's is strong and robust.

One stage up the ladder is Shure's VP64A which costs around £150 which again is a dynamic mic but one designed for radio and TV broadcast use. It is almost omnidirectional and will pick up sounds from the sides and back very well. The VP64AL version is a couple of inches longer and gives just that bit more 'reach' which is useful when interviewing.

Neumann

These are an excellent German manufacturer of some really first-class microphones, often seen in high end music recording studios and at some of the largest radio stations. The Rolls Royce of their range is undoubtedly the U87 which currently sells for just over £2,000, depending how many accessories you want with it. It's a large diaphragm with a double membrane capsule, three directional characteristics and a switchable low frequency roll off. It's ideal as a main microphone in just about every studio situation.

A much cheaper Neumann is their TLM67 which is the modern equivalent of their classic U67, which had a wonderful sound, due to its use of tubes (valves). The LTM designation is simply 'transformerless' which reduces the likelihood of electrical interference. The TLM67 microphone is renowned for its ability to handle high levels of sound and still maintain excellent quality. The TLM67 mic's cost around £1100   *https://amzn.to/47pHuHC*

A much cheaper Neumann microphone is the *TLM102 Nickel*, which has a large diaphragm with a cardioid directional characteristic. It is compact but can tolerate very high maximum sound pressure level. The unit has a slight presence boost above 6 kHz, adding a bit of sparkle to audio. It has transformerless circuitry and comes with a mic' clamp.

https://amzn.to/471HvkZ

ONLINE Radio

AKG

An Austrian company that has been producing top quality microphones and headphones for professional use since 1947. The company was founded by a physicist and an engineer are one of the best regarded manufacturers of microphones for radio broadcast use. They are designed and built in Vienna in Austria and most use their legendary one-inch edge-terminated large diaphragm. This is what gives AKGs such a great sound and an outstanding dynamic range (up to 156 dB SPL). AKG have a wide range of both dynamic and large diaphragm condenser microphones in their repertoire.

One of the most often seen AKG microphones in recent years was their D202. They had excellent quality and were very resilient to knocks and abuse. If you can find one 'used' they are worth picking up as they do have a reputation for not letting you down; they often come up on eBay. A cardioid dynamic microphone, the D202 has separate low and high frequency capsules with a switchable bass cut control.

They were often called 'the sound rocket' after their distinctive shape, seen almost everywhere in the BBC and still 'in vision' at the House of Commons. You can still hire these from many specialist audio suppliers, they charge around £24 a week which gives some idea of the esteem in which they are still held.

The AKG *Perception P120* is a modestly priced general purpose large diaphragm condenser microphone with a 2/3" capsule, a stainless-steel body. Its solidly built, gives outstanding performance and is excellent value for money. It's now available at just £77. It has a bass cut switch and a 20dB attenuation pad and a 2/3" capsule. Rugged die-cast housing and stainless-steel grille all serve to make this an ideal entry level mic.

The AKG C414 first appeared 45 years ago and they are still a strong seller with an excellent reputation, although its £800 price bracket may deter those with a smaller budget. The C414 has no less than nine different pickup patterns to select from which allows you to choose the perfect setting for every application. (there is 'lock' mode, so you can stop idle fingers spoiling your settings!). The choice of patterns includes omnidirectional, wide cardioid, cardioid, hyper-cardioid, figure eight and 4 intermediate settings.
https://amzn.to/48RYE1U

Peak Hold LEDs on the C414s detect even the shortest overload peaks and they have an incredible dynamic range of 152 dB. This 'engineer's choice' microphone has three switchable bass cut filters and three pre-attenuation

ONLINE Radio

levels. Leading-edge technology and state-of-the-art components ensure shortest signal path and it also has extra protection against moisture. Found in some of the very best radio and music recording studios, its often the choice of instrument for miking up acoustic guitars, etc.

AKG's C214 is a similar microphone that is currently selling for around £350 via Amazon which is more suitable for vocals. It has a cardioid response pattern and AKG's Legendary 1 inch Edge-Terminated Large Diaphragm. It has up to 156 dB dynamic range and Ultra-Low Noise. It has a switchable 20 dB Attenuation Pad and Low-Cut Switch. The rugged double mesh grill gives it high RF immunity but has little effect on the acoustics.

https://amzn.to/45BxXvB

Audio Technica

Often mistakenly referred to as *Technics* (who are another company altogether) *Audio Technica* are a highly regarded Japanese manufacturers of a wide range of audio equipment, particularly microphones, headphones and turntables. Their microphones are very robustly made and give great quality sound. The AT models most suitable for radio studios are these:

AT2020 a cardioid condenser microphone that is a side-address model with high SPL handling and wide dynamic range. Its custom-engineered low-mass diaphragm provides extended frequency response and great transient response. Needs a 48v phantom supply and does not have the necessary XLR lead in the box. The AT2020 usually costs £99.

https://amzn.to/3rXaibc

AKG also have a 2020USB model with similar characteristics and a high-quality A/D converter with 16-bit, 44.1/48 kHz sampling rate for superb audio, delivered through a USB connector. It has a socket to plug in an earpiece or headphones to enable you to monitor the output with zero delay on the sound. A mix control blends music with the mic's audio, making this almost a self-contained podcasting unit. Usually £169, but recently available for £104. *https://amzn.to/495SzyY*

World of Radio has acquired and purchased hundreds of microphones on behalf of radio station clients; there is a comprehensive list and partial review of microphones suitable for radio studio here.

https://worldofradio.co.uk/Microphones.html

Which microphone should I buy?

ONLINE Radio

There are several factors which need to be considered in this debate, not least of which are security, budget and what will the end user, your audience, be listening on. Most of them will not be able to tell the difference between a cheap 'used' mic that has been swung around by the vocalist in a band, or used and abused by hundreds or maybe thousands of drunken Karaoke contestants. There is a good article about '*Choosing a Microphone for Webinars'* on Jan Ozer's site, the Streaming Learning Centre.

https://bit.ly/3MmLrVn

A poor mic is better than no mic at all, but operators should get either personal usage experience or advice from a consultant before investing in microphones. These are often a key part of the radio station and the sound of them can lead to either long listening hours, or listener fatigue, which is very undesirable. There are some very good microphones in the 'bargain basement' department, it's impossible to lay down any 'hard or fast' rules as so much depends on what else you are connecting it up to.

Microphone processing is the key to get the best sound from any mic and there are more details on these later in this chapter. As with the microphones, getting to know the piece of kit and its capabilities is vital if you are to get the best from it. Always experiment and try out new ideas or advice you have heard.

Some consultants will have you spending a fortune on microphones, which are not guaranteed to make your radio station sound any better. Almost everyone in the radio broadcasting business is a closet anorak, and we have found that they are all keen to share their advice, anything for an anorak session.

A good property for radio studio microphone is the cardioid which rejects noises arriving "off-axis", great for self-drive operators who might be moving items and operating controls while talking, and probably shuffling papers.

ONLINE Radio

Microphone accessories
Of huge importance to the quality of signals from a microphone are the acoustics of the environment it operates. Adding acoustic treatment, usually in the form of sound deadening tiles to reduce audible reflections, will make a big difference and help focus the listeners ears on the speaker. Room surfaces are dealt with above, but some attention should also be given to large surfaces of equipment close to microphones.

Of great importance is the way in which the microphone is mounted. A radio studio microphone should ALWAYS be properly mounted, preferably from the ceiling but from a nearby wall or desking if necessary. The mounting arm needs to be adjustable so that it can be moved to an optimum position for the speaker.

An ideal swivel mount arm is the **RØDE PSA1** which is ideal for radio studio or podcasting. The PSA Microphone Mounting Arm mates with most microphone brackets. It has a good long reach horizontally and vertically and can be rotated through 360°. *https://amzn.to/3QoLwKD*

A good alternative might be the **EALLC** boom, which has the traditional scissor action and is a well damped. It's featured here on a section that's full of mic booms and other accessories. *https://amzn.to/3FrAQVb*

BOOMS
Radio broadcast microphones are usually held in place by shockproof mountings. These are invariably an array of elasticated cords, which will absorb and sudden jolts that the mounting arms experience, or the surface the arm is mounted on. The mount prevents shocks and jolts being transferred to the microphone. One of the most reliable microphone shock mounts that is designed to stand up to a lot of use in a radio studio is available_ for less than £35, with free delivery.

https://amzn.to/3tIQ3yl

Pop Shields
The front of the microphone would be protected against strong bursts of air from the speaker. They are most often heard when pronouncing the letters B, P and so on. You can tell which the sounds are by holding a lit candle in front of your lips; when making the plosive sounds, the flame will flicker, but not for other sounds you make. These explosive sounds often result in a popping sound, due to the short capsule of air rushing out as you make the sound.

ONLINE Radio

The sounds are emphasised by capacitor mics which have a very small and easily moved diaphragm. They can be easily reduced by the speaker or vocalist turning slightly off mic as they deliver strong bursts or words with known plosive bursts in them, but that's not always possible and it can mar a performance. These effects can however be eliminated by having a 'pop filter' in front of the microphone. Often called a 'pop shield' this can be a simple foam sock over the microphone, or a larger disc of dual layer gauze mounted on an adjustable arm and clamped to the body of the microphone. The fine mesh of the filter breaks up the initial rush of air and lose its turbulence. There are dozens available at:

https://amzn.to/3M44HH2

The placing of a pop shield is very important, which is why the shields on an arm work better than a simple sock on the microphone, although even that would be better than nothing at all. The shield should always be at least two and up to four inches in front of the microphone.

Keeping a reasonable space between the shield and the microphone also serves to keep the speaker some distance back from the microphone, which not only gets better reproduction but is good for hygiene and protects the microphone. Nylon mesh pop shields can cause some dulling of treble frequencies, but metal mesh designs not so much.

MIC PROCESSORS

While you can simply plug a microphone in to a mixing desk and use general equalisation controls to reduce sibilance and other undesirable effects, there are a lot better ways of modifying a microphone's output to achieve a really good sound. Microphone signals are at very low levels and need boosting by a pre-amplifier before being manipulated. This is done to improve the 'signal to noise ratio' which is the difference between the audio and the background noise. That hiss is often called the 'noise floor' or ambient noise level.

Very basic processing can be done by passing the mic signals through its own graphic equaliser, which will reduce or boost bass or treble frequencies as required, and may have enough channels or filters to notch out or boost certain narrow bands of frequencies. This can give an overall improvement to the sound quality and intelligibility. By gating out quiet bits, between the 'wanted' speech, a voice can be made really punchy.

ONLINE Radio

Any microphone pre-amp should be as early as possible in the circuit, and should be 'transparent'; i.e. it doesn't change the sound at all, just boosts it. Behringer have a very simple four channel mic preamp (their PP400, costing £22) that converts mic level signals up to 'line level'. Other audio sources, be they grams, tape recorder or whatever, are likely to be feeding a radio mixing desk at 'line level' which will be around 1 volt for 0dBu – a reference level that goes back to the earliest days of telephony and is referenced to the 600 Ohms line measurement.

https://amzn.to/47pakl3

No two voices are ever the same and it's important to individually stylise the sound to match the speaker and how you want the end result to sound. Careful application of microphone processing will improve the sound as well as the consistency of the many varied voices heard on a radio station.

Some microphone processors can add 'tube pre-amp' effects to the signal. Traditional tube (valve) circuits tended to emphasise the even harmonics of the voice, (2,4, 6 times, etc) whereas as solid date pre-amps will favour the odd harmonics, (thirds, fifth, etc). This is what adds colour and perceived warmth to a valve circuit.

Other circuits that are often found in audio processing include expanders (which increases the gain in quiet periods of speech) limiters, which keep the maximum audio levels down to a certain level, de-essers and equalisation. Each one of these sections of a microphone processor has their own advantages, and their use is to some extent personal taste. It's important however to read the instructions that come with a processor to learn exactly what each circuit does.

A gated compressor set up right can add a lot of energy and excitement to a voice that might otherwise sound just a bit 'flat'. This is very important if you want the dialogue in your radio station to be heard and stand out. It needs to sound at least as good as the processing that the engineers at recording studios will be adding to the music you are playing, so it sounds to be a part of it, or congruent to it.

World of Radio has a list of microphone processors they've used at radio stations and a selection appear at the link below. They also have a comprehensive list and partial review of microphones suitable for radio studio use currently available in the UK here.

https://worldofradio.co.uk/Microphones.html

BEHRINGER

One of the cheapest microphone processors, and certainly the most effective, you will find is the **Behringer ADA8200**. The uprated edition of the 8000 Ultimate Digital, it is one of Behringer's best-selling products. The ADA 8200 is easy to operate, fits neatly into a single rack space. It has MIDAS-designed mic preamps and integrated analogue to digital converters. They give optimal signal conversion without distortion.

This new model has advanced Cirrus Logic converters and MIDAS-designed Mic Preamps, making it an excellent choice for musicians and audio engineers alike. Frequency range is an ultra-wide 10 Hz - 24 kHz at a 48 kHz sampling rate. Connections are provided for balanced XLR and 1/4" TRS, as well as unbalanced 1/4" jack inputs. XLR line out connectors on the rear panel gives either balanced or unbalanced output. Digital input and output are handled via ADAT format TOSLINK optical sockets, which can support up to 8 channels of data. This level of control over your audio would have cost about £1,000 a few years ago but the Ultimate Digital can now put one in your studio for just £175.

A good two channel processor is the **Monacor** Stage Line 2 channel compressor limiter which has noise gating and side chain input. Often used in studios for transferring analogue recordings from vinyl or tape into digital, where it's often used to make the sound much denser.

Monacor's Ash has LED indicators to help adjust threshold values for expander, gate, compressor and limiter. It has a switch for hard or soft compression (soft knee) threshold control (for limiter too) and rate control. Threshold, ratio, attack, release and output level controls for compressor. Automatic mode attack and release for compressor.

dBx 286S

The dBx 286S is a very versatile microphone pre-amp with compressor, de-esser, LF & HF detail & expander and gate. It provides all the mic' processing tools you need in one box including a full featured processor with preamplifier and four processors that can be used independently. It has wide-ranging input gain control and an 80Hz high-pass filter to remove low frequency hum, rumble or wind. These artefacts are best removed BEFORE your regular audio chain. Consider dBx's *OverEasy* compressor to transparently smooth out uneven acoustic tracks, It eliminates vocal sibilance too! Usually sold at £180 they are often on sale at Amazon for around £155.

https://amzn.to/491vygB

ONLINE Radio

Processors for radio station output

Before leaving the radio station studios and being sent to the server, it's important to have the audio as closely matched and processed as possible to how you want it to sound. Radio stations previously would have a small army of engineers monitoring the station output, constantly making slight adjustments to the audio. These were primarily for equalisation. A modern piece of circuitry handles all that automatically and adds compression and expansion where necessary. The really professional kit, such as Bob Orban's excellent range of *Optimod* processors can cost up to £6,000, but some quite good ones can be had for as little as £250.

The secret to good audio fidelity is to not over-compress it to within a tenth of dB as this adds distortion which leads to listener fatigue. Increased apparent loudness by narrowing the dynamic range which can make a station sound stronger and make it stand out but it neds "good ears".

There is a full section on processors with a range of options on the World of Radio website. These are items that are not usually available "off the shelf" in the UK and are usually constructed to order – allow a couple of weeks for delivery - unless you want to pay a premium price.

https://worldofradio.co.uk/Processors.html

There are several models of processor that have been specially designed for streaming (Online Radio) use:

The **ORBAN PCn-1600** is a software-based (Windows) processor capable of handling multiple channels on one device. It can handle signals for DAB+, streaming radio and podcasts.

Orban's exclusive *PreCodeTM* technology manipulates several aspects of the audio to minimize artifacts caused by low bitrate codecs, ensuring consistent loudness and texture from one source to the next. PreCode includes special audio band detection algorithms that are energy and spectrum aware. A selection of factory presets help to find the perfect sound. A PC with Intel Core 2 (3GHz or faster) is needed to run it.

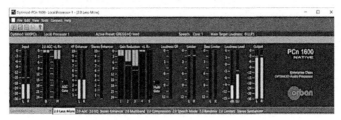

ONLINE Radio

Compact in its design but big in functionality, the **OPTIMOD Trio** can be configured for use with traditional AM or FM transmitters, or alternatively for DAB+ and streaming output -Online Radio. The Trio is even suitable for TV audio processing! It is probably the most flexible processor and one that can always be relied on if a station's audio processing requirements change.

The Omnia A/XE features adjustable wide-band AGC with a three-band compressor and limiter, a low-pass filter, and a precision look-ahead final limiter to prevent clipping. Stations will sound cleaner, clearer, and with more presence and detail.

The Omnia A/XE is a market-dominating, competition-crushing beast that sounds so pure, clean and compelling that it's almost an unfair advantage. It is capable of processing audio for a wide variety of applications, both compressed and linear. It runs as a Windows service, can be fully-managed and configured remotely using your favourite web browser, and can even process and encode multiple streams in various formats simultaneously. With the Omnia A/XE you can encode directly to MP3 or AAC, feed a Shoutcast-style server in the MP3 or stream to Adobe Flash clients through a Wowza Media Server. You can also link the Omnia A/XE with your existing Windows Media, Real, *mpgPRO* or MP3 streaming encoder.

A new *Virtual Patch Cable* allows the Omnia A/XE to receive, process, and send audio to other software on the PC. Shoutcast or Wowza server streams can be tagged with now-playing information received from any automation.

The Omnia F/XE is a file streaming audio processor and encoder that is probably the most advanced file-based application available. This model combines Omnia audio processing with both the MP3 and AAC codecs for high quality file preparation for podcasting or online radio.

The F/XE uses Omnia's special processing circuits to improve loudness and perceived quality. It is software only, so special cards are not required. It is able to read PCM WAV files, both Layer-2 and Layer-3 MPEG files and it can automatically send the output file to an FTP server. It will also notify you by email if problems are detected and keep logs of where any problems have arisen.

ONLINE Radio

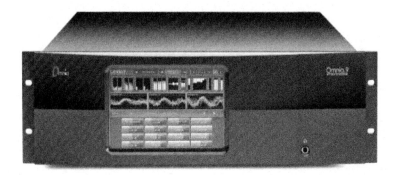

Omnia

The larger **Omnia 9/XE** is the top of the range, and can remove distortion while selectively undoing the over-compression that is so often found in commercial recording. It optimizes sound quality of music that has been reduced to very low bit rates (read 'quality'). The 9/XE does this by removing the many distortion components so that they you are not waste data bits during the encoding process.

It has a 6-BAND parametric equaliser and three-stage AGC, with adjustable sidechain filter for consistent levels. It supports multiple ASIO and WDM (Wave/Direct- Sound/Kernel Streaming) audio interfaces simultaneously. A flexible remote-control system with touch screen support, comprehensive monitoring, and remote audio streaming are all included.

TEKO Broadcast, and **AxelTech** are two companies offering processors from their bases in Italy. The *Falcon* range is lower priced than the usual high end processors found in the American market, but we have been unable to find sufficient information on them to recommend.

WAVEPAD is an audio editing software package that woks on the Mac or Windows operating systems, and allows users to then add effects like echo, amplification and noise reduction. WavePad works as a WAV or MP3 editor, but it also supports a number of other file formats including VOX, GSM, WMA, real audio, AU, AIF, FLAC, OGG, and more. It also includes a library of sound effects and many other features.

ONLINE Radio

Headphones

There are many types of headset available. Those found in radio stations seem to have had a very hard life, so it's best to get headphones that are designed for professional use and can stand for some hard knocks. In its lifetime that poor headset will adorn dozens of heads, be torn off in frustration more than a few times. Worse: a pair of headphones will invariably get run over by the studio chair, quite a few times too!

Professional operators might wear a headset for four or five hours, so it's vital that these are comfortable, they fit well and they are the right type. A lightweight but 'broad footprint' headband is important for comfort and the size of the cups and the cushion material used is probably most important of all.

DJs and producers who work in night clubs are very aware of the importance of getting the right headset – one that they can actually hear properly in environments that can be aurally challenging, or even hostile. AKG now manufacture special TIËSTO Signature headphones that give superb results. *https://amzn.to/3Qlf9wc*

Among the features of headphones which may make up your mind are foldability, which make it much easier for them to transport or pack away. The BBC studio headphones have always been of questionable quality and most radio presenters prefer to bring their own headsets.

Some headsets can be specially configured to accept two channels of audio; not simple stereo, but separate channels for cueing or auditioning, or a talkback line. The other channel can be configured to hear the output of the desk, or even the on-air sound of the radio station.

Among the best headsets, which last longest and give a good wearer experience, is the *Beyer Dynamics DT770*. A closed cup, dynamic model, they have bass reflex technology for improved deeper sound response and a comfortable fit thanks to its adjustable, padded headband construction. Fairly robust, with all parts replaceable, including the ear cushions.

https://amzn.to/46BLNzw

ONLINE Radio

Numark

Numark make good quality headphones and their model are very popular in radio. Their Red Wave headphones were designed with DJs in mind. Their superior sound, eye-catching looks and comfortable design make them so popular with professional club DJs. Their closed-cup design gives good isolation.

Large 50 mm neodymium drivers produce clean, powerful output and they have a swivelling design on one cup so they're easy to wear one-sided, between the ear and shoulder. The ear cups also have breathable protein-leather which helps them stay comfy on long shifts. ***Gear4Music*** have around a hundred models of headphones available.   *http://tidd.ly/9fbd1a0*

Sennheiser's 'Basic Edition' model, the HD25, has a split headband which supposedly gives a more secure placing on the head. They are very sensitive, thanks to aluminium coils in the speakers and give good attenuation of background noise. They are usually sold for up to £180 but recent price wars have seen them on sale at Gear4Music at £149.

Gear4Music also offer an optional five year warranty   *http://tidd.ly/432974a4* while Amazon offer the same cans for £159.   *https://amzn.to/3QI5LHF*

Durable and comfortable, these have been the radio industry leader for about thirty years now, thanks in part to a two-year guarantee. That's well beyond the usual life of a set of 'lesser' headphones in a radio studio. The HD25s were the headsets given to passengers on British Airways' Concorde flights and those models with the 'Concorde' logo on the headband have now become very collectable among DJs.

Sennheiser also offer headsets with more conventional broad headband such as the HD. These are a 'Professional Producer' headset with superior comfort for long air shifts and very rugged build. The proprietary Sennheiser acoustics system gives you an accurate and balanced sound needed for mixing and monitoring.

ONLINE Radio

In Ear Monitoring

In-ear 'buds' have become very popular recently and are not only seen and used for TV interviewers getting last minute instructions from their producer, but worn by singers and rock stars on stage. They come in a variety of 'qualities' from the Shure SE846s, (read a full technical description here) which cost an eye watering £799 which are available for around £900 from Gear4Music, through their SE535 which are a more affordable £320.

ARTICLE : *https://cdns3.gear4music.com/media/7/76920/download_76920.pdf*

In-ear types are fully sound isolating with excellent quality true sub-woofers giving extended clarity. Shure also make a SE2215 at just £78 – the details are at this link: *https://amzn.to/45GIFSD*

AKG have a model called the K3003, a fully professional 3 way in-ear bud, with an integrated microphone. At £105 <u>a set</u> they are not the cheapest, but they are described as the best on the market and have anti-allergenic sleeves and great with Apple kit.

https://amzn.to/3McOq2m

Even those are not the most expensive. Just a few quid short of £9,000 will get you a pair of *Happy Plugs* in 18 carat Solid Gold. Hand-crafted by a Swedish goldsmith, they have 25 grams of solid gold. Probably the ultimate combination of exclusive jewellery and high-tech electronic accessories, they do bring a whole new meaning to the words "ultimate luxury finish" in what is one of the smallest everyday items.

The Philips white buds take in-ear technology one step further and offer oval sound tubes, BlueTooth, and enlarged 13.5 drivers for a rich, deep, bass sound. They are sensitive right down to minus 108dB they weigh only 14g (less than half an ounce). *https://amzn.to/3QI5LHF*

ONLINE Radio

The Sennheiser earpieces are much cheaper (the CX300 are only £26.45) and almost all give great clarity of sound thanks to bass-driven speaker systems. The ear pieces are a personalised fit in the ear canal. We recommend that you avoid the really cheaper ones however as these can have bad sound peaks; you don't realise you are subjecting your ears to 'bad sound' until you find you have gone slightly deaf and are experiencing tinnitus and other problems.

Audio delay and latency
Beware. Some BlueTooth transmitters (that send your audio to the headset) have far too much 'latency' or delay to the audio. It can be very disconcerting to hear the time lag between actually speaking and hearing your voice back in your ears. As little as 50 milliseconds is too much for some people and causes lack of concentration, and faltering speech. Many digital circuits will not be able to correct this and it's important to choose a professional "designed and fit for the purpose" combination, get some good advice and even better, try it out first.

Professional stage kit, as used by musicians when performing live, uses a range of techniques to avoid latency problem.

The **Zuwit** system uses a spot UHF frequency (around 433 MHz, which is adjustable, to avoid interference) and has a comfortable earpiece, a charger and has been a best seller for about five years. Their Wireless in-ear system has six selectable channels and an impressive range and usually costs just under £60.

Gear4Music also have an excellent reputation for this kind of equipment, with excellent after sales service. They sell over fifty in-ear monitors many of which can be seen by clicking this link.

https://bit.ly/46CkvZN

ONLINE Radio

Consoles

The heart of a radio station, as far as presentation goes, is the mixing desk, often shortened to simply "the mixer". It is also variously called a console, a board and so on in different countries. This controls the various programme sources and adds them together before sending them to the radio station's RACKS - usually a small room full of equipment mounted in metal racks, with easy access to the backs and front of the equipment and normally only inhabited by engineers.

Mixers are generally problem free until a presenter decides to feed them with a drink or other substances. Rotary faders were most 'presenter proof' as there was little room for the tea, coffee or coke to get in, and being mounted almost vertically there was less chance of a heavy pile of vinyl or anything else being dropped on them. Since horizontally-mounted linear faders became common-place in the 1970s, the mixing desk has become much more vulnerable. Some studios introduced a firm 'no liquids' policy but the desk is still the most vulnerable item of equipment in a studio.

D & R

D&R's Airlite mark2 is virtually a complete radio system, ideal for ON-AIR, Production and especially streaming to the web. It is built into a heavy-duty RF shielded chassis, designed with elegantly rounded corners. It has 3 balanced, phantom powered mic inputs with inserts for voice processors, 8 stereo line inputs, 4 USB channels and 1 Telephone VoIP channel.

ONLINE Radio

The most 'up to date' mixing consoles have touch screen. A high-resolution monitor, it has the entire surface given over to the controls of a virtual radio studio, with the couple of dozen controls needed to run a radio programme are presented in whatever size is convenient. All but the most important controls can be hidden at whim, with items such as EQ and SFX which are not in constant use. Every control can be operated simultaneously, depending on the dexterity of the operator and there is no audio at the screen, everything is safe in the racks area. The console is simply a huge remote control.

The move to digital mixing, where the desk is simply controlling a remote piece of audio equipment in a racks room elsewhere, is a tremendous boost to reliability.

At its simplest a mixing desk is just a set of controls that will mix together the outputs of your sound sources. The complexity of the mixing console needed will depend on the type of programme you intend having, in particular how many sources will you use, and how much control will the operator have over the playout process.

The mixing desks seen in recording studios will often have many channels into which the sources can be fed. 64 and 128 channels is becoming standard, and each of those channels can have up to a dozen separate controls with which to set various equalisation parameters, sending to effects such as reverb' and so on.

As recording studio desks have become more complex, the trend is to have many of the controls automated, with memory settings for optimum levels of many variables. A recording studio desk however is neither necessary nor indeed optimum for radio broadcast use. The main function of the mixing desk, also known as a console, is to control events and audio levels. It's a switchboard with additional controls for varying audio levels and for equalising the various sonic qualities of the signal. The master control room as it was once called, is simply a control hub.

ONLINE Radio

Control hub

The mixer is the heart of a radio station and as such is not simply a tool for varying the sound levels or EQ. It should also be the place where all the various programme sources are controlled; switching sources in an out of circuits, remote starting items and so on. Sources need to be 'auditioned' or cued up ready for airing, some need to be sent to external processors, or to be recorded. The needs of a radio station mixer are many and various and quite unique to mixers found in recording studios, or in a live environment, such as on stage.

At the very least, a radio mixing desk needs to have a method of cueing, or auditioning a source, without putting it into the programme chain (live on the air). Cue circuits (usually a separate channel) enable the precise start point of the source's programme material to be found and then handed to a remote start control on the desk. This is usually achieved by having a micro-switch connected to a volume control, of either the rotary or linear(slider) type. As soon as the control is advanced it triggers the micro-switch, starting the event – usually a recording. 'Instant start' of recordings sounds much slicker, with a seamless playout and no embarrassing gaps between items.

When connected to a studio microphone the switch often triggers a red 'on air' light to illuminate. Radio desks often include a talkback circuit so the operator can converse with others in adjacent studios, or in the announce booth. The audio mixing desks used on stage or in a recording studio For the very latest details of audio consoles and mixing desks that are suitable for use in radio stations, plus further links to specific equipment, have a look at the web page about this.

https://worldofradio.co.uk/Mixers.html

Both *Behringer* and *Sound lab* produce a wide range of small mixers that are designed for use by vocalists and musicians, but which are available cheaply and could be pressed into use in radio. Some are shown here, with the latest prices. They will handle the audio and equalise it well, although some find the controls fiddly.

https://amzn.to/3Q5lc5y

ONLINE Radio

Arrakis Systems

An American manufacturer of mixing consoles with a wide range of capabilities. Included in their repertoire are desks designed with Internet Radio stations in mind and a special internet streaming service made simple.

The ARC-8 radio console is an eight-channel single output mixer that's compact and durable. It's made for professional radio applications using a strong steel chassis. It has linear faders, switches with LED lit indicators and it has a Bluetooth input so you can feed in audio from either a phone or a Tablet. There is a phone channel with mix minus and a Talkback button so the DJ can speak to callers off the air. It has two high performance mic channels and a USB channel to connect to a PC or a Mac for playback or record. This console will do everything you need to run a radio programme and it costs about £750. You can see the ARC8 demonstrated in a video on their webpage or read the full specifications.

http://www.arrakis-systems.com/arrakis---arc-8.html

Axia Audio are the studio and audio part of the Telos Alliance. In 2003 they invented 'audio over IP' for broadcasting developing broadcast audio equipment for transmission over standard Ethernet. The mainstay of their technology is **Livewire +,** a patented AoIP protocol. By using everyday Ethernet cable and connectors, the cost and reliability of AoIP have been transformed.

Fusion is Axia's new modular console and it has heaps of features; Clark Novak is the best person to run through all its attributes and he can be seen on this video doing just that. They are a very attractive looking console, with lots of LEDs and specially constructed faders that repel liquids.

https://youtu.be/AasoW2Mb578

Axia's **Element** console is popular around the world; there are now over 5,500 of them out there as the main workhorses in thousands of radio and TV stations. It not only has four stereo buses, but it also features digital EQ, dynamic mic processing and such initiatives as virtual faders which let you run five channels at the same time over a single fader.

ONLINE Radio

D&R

A well-known Dutch company formed by Duco de Rijk and Ronnie Goene in the early 1970s is **D&R**. Duco had a number one hit in Holland with his band's superb version of the lead song in the show 'Hair', thanks to lots of airplay on Radio Veronica. D&R have also built desks for many famous Dutch music artists for stage use and the main studios of Radio Luxembourg too.

Their equipment has unmatched 'headroom' giving superb sound. Their mixers have a tremendous fan base, due to the quality of the electronics and the structure. R&D still supply spares for many of their products going back forty years. The main models of interest to radio stations are the **Airlite**, the **Airmate**, the **Airence** and the **Airlab**. All DnR models are available in the UK.

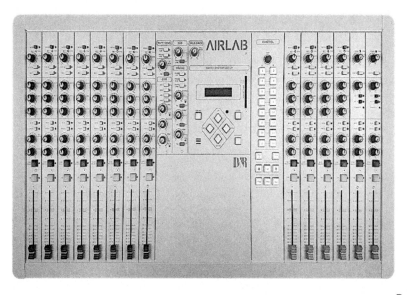

D & R

The Airlab radio console is extremely reliable and simple to operate. Its analogue, digitally controlled signal path giving the best solution of both analogue and digital worlds.

Full Spec & details: https://dnrbroadcast.com/

Latest UK Prices: https://amzn.to/3McNdZc

Lawo

ONLINE Radio

One of the most innovative manufacturers of radio station equipment in recent years is this German based company, founded in 1970 as an engineer's office. Lawo now have around 200 employees in North America, China and Switzerland. Their equipment is very popular in theatre and live presentations, in broadcast (radio and TV) production and in 'on the air' radio studios.

Lawo's main audio production studio is called Jade and is very versatile. The *SapphireCompact* is a good 'on air' console, ideal for radio studios featuring Automix, an intelligent DSP algorithm, multi-layering and a flexible control surface. It's all powered by a 1RU engine which not only powers the controls but has signal processing. At the heart of the Lawo's digital consoles and routers systems is the unique Universal Networked Audio Engine, which has a stand-alone Micro Core. The Lawo systems run on Windows 7 and above.

The **Lawo CrystalClear** console is faster and easier to use than conventional boards and has a variety of advantages, all set out on Lawo's website. Unencumbered by physical knobs, switches and faders presenters are less intimidated and can focus more easily on their programme's content. Just touch it a few times and you will be convinced.

RƎLAY Virtual Radio mixer
Lawo's VRX is the world's first true Virtual Radio Mixer. Not just a screen controlling hardware in a rack, this is a real broadcast mixer that's *completely software based*. Mixing, processing, routing; all at your fingertips. The RƎLAY can mix audio using a PC.   *www.lawo.com*

Lawo Inc

ONLINE Radio

SONIFEX

These are a long-established British company, based in Northamptonshire and have long manufactured a range of equipment for radio stations. They have been making audio equipment since the 1980s. Their basic mixing desk, is the S0, a high quality but easy to operate unit. It's very suitable for Internet Radio stations and even includes a telephone hybrid for taking and receiving calls from guests or listeners live on the air. It has sockets for plugging in MP3 players and even a USB connection for interfacing with an automation system via a computer.

The headphone feeds on this mixer have limiters included and the main microphone control automatically mutes the speaker monitor section. There is a large LED display for monitoring levels and both programme and record outputs. The mixer can be rack mounted or set flush into a desk

The S1 was Sonifex's intermediate console with a proper broadcast solution without compromising on quality. Excellent mic amps, comprehensive facilities & simple, reliable operation. It is a compact, low cost, fixed format mixing console designed for on-air radio use.

The Sonifex S2 is a modular mixer coupling digital audio quality with analogue reliability. It accepts both digital and analogue inputs, with simultaneous analogue and digital outputs. The S2 has PFL/cue, fader-start operation of sources, automatic muting of monitor speakers on mic going live, controls for 'studio live' lights, EQ on input modules, gram amp input options and twin audio buses. You can see the full specs of the S2 here:
https://www.sonifex.co.uk/s2/index.shtml

Sonifex S2 mixer

Sonifex Ltd

Chapter 6

FORMATS

A radio format describes the overall programme output of a station, in particular the music genre played and, to some extent, who the radio station's target audience is. Radio has slightly different terms to describe the various genres of radio formats. The main ones which apply particularly in the business of Internet Radio broadcasting are highlighted further on in this chapter.

Not every item of music will strictly fall within the precise description of music, but overall, listeners will expect items to be stylistically consistent with a station's chosen overall sound. It is important that radio station operators and members of their team, understand:

(a) what a station is trying to achieve in terms of listenership; and
(b) what the components named in its format mean.

ONLINE Radio

What a radio station's music format actually sounds like is set by four main parameters; its style, the time period, the activity level, and the level of sophistication:

Music Style refers strictly to the type of music played, regardless of how the music is packaged for airplay.

Music Period refers to when the music is released. 'Current' music generally refers to that released in the last few months, 'Contemporary' music generally refers to music released within the past 15 to 20 years, or a generation. 'Oldies' is music released between 20 and 50 years ago, while 'Nostalgia' tunes are over 50 years old.

Music Activity A measure of the music's dynamic impact, ranging from soft & mellow to loud & aggressive. Some names of music styles incorporate descriptions of the music's activity level. Such as 'hard rock' and 'smooth jazz'.

Sophistication is a reflection of whether the musical structure and lyrical content of the music played is simple or complex. It is reflected in its presentation on air and often drives the composition of the audience.

Performance is any public use of a song or piece of music. The performance can be a live rendition of the work, or someone playing a recording of a performance, such as an album or a downloaded track.

The **owner of a copyright** has five distinct rights, which include copying, adapting the work, distribution of copies and public performance. The owner can sub-let those rights to anyone else, by way of a licence.

ONLINE Radio

STATION NAMES

The format of a station and its branding should be reflected in its name, which ought to tell listeners what to expect. If the name is in any way vague or oblique and doesn't immediately suggest the format, then the station's tag line or positioning statement is vital. Stations should never waste this opportunity to tell prospective listeners what a station is all about. Bland meaningless slogans, such as "More music, More hits" are pointless; more than who exactly? Most stations operate around the clock and play roughly the same amount of music. This applies to almost any type of radio station, whether it be a traditional over the air or one delivered wholly online.

Station names are important, so operators should choose one that's unique and is memorable. It might intrigue listeners and pique their interest if its mysterious enough, but few have spare time to try and understand a station's ID if it's obscure. Never choose a name of another station just because there is a convenient set of jingles available (which may be copyrighted – beware!). Using an old station's names is unoriginal and confusing but it can be misleading and annoy listeners.

Geographical hints in a station's name might help listeners identify with local or neighbourhood stations, but can be constricting if a station plans to expand. Careful thought should be given to the promotional possibilities. It costs money, listeners and credibility to change a radio station's name.

There is considerable reuse of names as one would expect. It was never a big problem when most stations were only audible within their own counties, or even a small local area, but online radio changes everything.

Stations are often called after the place of original, either of the studios or the transmission site. The really wise ones choose an area that easily described and encompasses the target area with which listeners associate. Examples are *Britain Radio* and *Radio England* both active in the sixties, or the many BBC 'national' stations *Radio Ulster, Radio Wales, Radio Scotland, Radio Cymru, R Nan Gaidheal*, etc – but odd that they don't deem England to be worthy of its own 'national' voice, even though there is a separate *BBC Asian* channel. When the BBC ran local stations, these would often be named after major cities, but they later used county names, even when the new counties were hated by the locals. The corporation's march to regionalisation suggests that this was a long-held intention. For over fifty years, *Radio Luxembourg* broadcast from the Grand Duchy in the heart of Europe and made the tiny country known to millions; amazing promotion.
The most effective names however are probably those that are most appropriate; i.e., the radio stations that sounds like and does exactly what it

ONLINE Radio

says on the tin. Many of these are extensions of a radio stations programme format – such as. *(county name) Sport*, or *(city name) Rock*.

This is of increasing importance now that radio stations are not confined to one small geographic area. Many stations that were once confined to a city or small region have now added their output to the internet and so can be heard globally, even though their output might not be of worldwide interest.

Traditionally, some stations would be known as *Radio XXXX* after the name of a town, city or region. Later, the word 'radio' was dropped as everyone knew it was radio as that's how they tuned into it! Stations began using the word *'sound'* as an appendix to their place name.

In some countries the station's licence was a call sign with a seemingly random group of letters, such as WABC in New York or WGN in Chicago, which stood for *World Greatest Newspaper* and was owned by the Chicago Tribune. Many newspapers were the pioneers of radio broadcasting; in Britain, the *Daily Mail* was among the pioneers of British radio and even funded a ship to tour holiday resorts and publicise the newspaper and radio.

Among the single word names that have long been popular to radio stations in many countries are the following, which are heard on different stations in many places and all unconnected to each other:

Arrow.	Atlantis	Bay	Beach	Buzz
Chill	Classic	Coast	Cool	Dream
Drive	Energy	Fire.	Flame	Fresh
Funky	Gold	Groove	Heat	Heart
Hot Hits	Imagine	Joy	Jungle	KISS
Laser	Melody	Metro	More	Kickin'
Nova	Original	Pirate	Quality	Revive
Sky	Smooth	Star	Sunshine	Super
Talk	Tempo	Viking	Winners	Zoo

Adding a radio dial slot to a single name was popular for many years and eventually many simply added a strident sounding letter, giving rise to Arrow-93, 96-X, Kiss 100, Y-100 and Laser 558.

Radio Veronica, whose name came from the company's initials VRON, began a trend of using girl's names, followed by Caroline, Jackie and Susy.

ONLINE Radio

More original stations select a unique name for their station, such as *Sound of Spitfire* which is based in Spalding in Lincolnshire. Apart from the roar of those mighty Merlin engines in its top of the hour identification announcements, the Sound of Spitfire programmes don't seem to have any other links to the planes, though the station is led by an air-traffic controller, Chris Hyde. It's aimed at audiences globally and has a widely-spread audience for its esoteric mix of a wide range of music, which includes lots of mod, ska and gold music, as well as shows of laid-back music, jazz and blues along with a sprinkling of new independent artistes on specialist labels. SoS encourages listeners to participate by using a Messenger group called *Goldmine,* that's named after one of Spitfire's most popular late-night programmes.

https://SoundOfSpitfire.co.uk

Cornucopia of Names
An online blogger called Brett Lindenberg recently collated a register of over 500 'clever' names being used by radio stations, most of them in the USA. Many of the suggestions can be found on his *FoodTruckEmpire* blog (not such a good name for radio!) and most revolve around descriptions of programme types, such as *Rocking Radio* or *Turn it Up*. Even more are simply a buzz word for one radio format or another, such as *HipHop Radio, Symphony* and one called *Radio Jingle*. Among the stranger suggestions are: *Message Received, No Requests Granted* and *We Don't Stay Silent*.

Brett's blog with radio station names: *https://tinyurl.com/yscwyu2o*

BRANDING. An incredibly important for any media operation and radio stations are no different. It's vital to make the radio station standout from the rest of the stations audible, especially in a market where there are over 100,000 others that your listeners can choose from. Having a catchy and memorable name can make a station really stand out.

A suitable domain name should be an early purchase as the good ones do go very early. To maximise impact, a radio station should have its identity as part of its domain. Having a memorable station ID that's also the website domain makes it easy to find and generate traffic. This applies to email addresses; keeping them uniform and similar to other social media IDs avoids confusion and will result in higher numbers of visits.

ONLINE Radio

Common Radio Formats

There are over 80 basic formats used in radio broadcasting. Most describe the varying output of music-led stations and they have evolved over several years as radio has become increasingly diversified, especially since the advent of Online Radio. Many formats have sub-genres too.

AAA   This stands for Adult Album Alternative, a format which is designed to appeal to adults and those with a more developed music taste. AAA stations tend to play more album tracks rather than hit singles.

ACTIVE ROCK   This term is used for stations which play mostly hard rock, metal rock and 'heavy metal' (see below).

ADULT ALTERNATIVE   is a format of appeal to adults with more developed tastes but usually playing only current hits.

ADULT CONTEMPORARY   AC stations play music from the past ten years or so, maybe twenty, rather than the latest hits. They are 'today's' artistes, but usually their better-known recent hits of the last decade or so.

ADULT ORIENTED ROCK   A term not often used today but in the 70s it described stations that played rock album cuts rather than singles.

ALTERNATIVE ROCK Originally this applied to punk music; which were originally mainly 'thrash' bands who played with great energy and gusto and invariably a lack of melody. In the 90s it began to be applied to rock influenced Top 40 'Brit' bands too. It appeals to the younger demographics and includes current hits and cuts from recent albums.

AMERICANA This genre has overtly American lifestyle music and includes country-rock, folk-rock, blues and American roots music. Appeals to adults more than to teenagers.

BIG BAND A music industry term borrowed by radio to describe the orchestra-led music that was popular in the 1940s and 1950s. These huge ensembles were usually fronted by well-known leaders such as Count Basie, Glen Miller and Nelson Riddle. Most had a lead singer who was of lesser importance than the orchestra or the band leader. The style was usually upbeat as their main work was in dance halls; a good example of Big Band music would be Frank Sinatra's "New York, New York".
  BLUEBEAT   The word Bluebeat has become a generic term for early Jamaican music thanks to a record label set up by Emil Shallit (Melodisc)

ONLINE Radio

which released R&B and pop hits from Jamaica. The name was coined by label manager Siggy Jackson, who explains "It sounds like the blues and it has beat'. Several unlicensed radio stations have used Bluebeat in the British Midlands and the north over the years and the term remains popular with Mods.

BLUES Music originally from the 'deep south' of the USA, fusing African and American folk music, often sad, from 'feeling blue' it's often based on trance like rhythms known as 'the Groove'. See Also R&B (below)

BOY BAND Vocal harmonies from young male singers, usually backed by orchestras or session musicians. Best known artistes are The Bachelors and the Walker Brothers (1960s) The Osmonds in the 1970s, Bros in the 1980s and more recently Westlife, Boyzone, Take That and One Direction.

CHR It traditionally stands for 'Contemporary Hit Radio' and is a term first used in the late 1970s by 'Radio & Records' (now part of Billboard magazine, the leading American music business paper). The term was coined to describe stations previously known as 'Top 40' which majored on current chart hits. As artistes began to enjoy longer periods of popularity, the description 'contemporary' has now become much wider.

Ten years or so can be regarded as contemporary and consequently, the CHR letters have often become regarded as Current Hit Radio. Many CHR stations focus on particular styles of music, and the term CHR has many sub-genres, such as CHR-Urban, CHR-Country, CHR-pop, etc.

CHRISTIAN Popular music that promotes Christianity. Many sub genres exist; Christian Rock and Christian Gospel being the best known.

CHRISTMAS
Stations playing lots of Christmas music, usually aimed at retail and restaurant operators to play in their business, often in the background. Playlists can be very limited, one UK Christmas channel offered only 32 numbers just a few years ago (perhaps they had programmed only part of a *'NOW its Christmas'* album!). In contrast, Sky Radio's Christmas channel programmed over 500 songs. With over 1700 Christmas songs now commercially issued and available, not including Christmas carols, there is no excuse for a limited playlist.

CLASSIC ROCK Featuring mainly the best-known rock music hits of sixties, seventies and eighties, by major artistes. These are usually both

ONLINE Radio

singles and album cuts. The format is known in two American cities as 'Rock Favourites'.

CLASSICAL Not so popular as a radio format as consumption usually demands very close listening, but it is highly regarded by some. Intellectually it has the high ground in music. Tracks often have lengthy complex arrangements and is often revived or plagiarised into contemporary music. Classical music is invariably instrumental.

COUNTRY One of the most popular music formats in the USA, it's based on traditional folk music of north American white settlers, particularly from the rural south areas. Vocally led, it features traditional instruments such as fiddles, banjos and pedal steel guitars. Often derided as cowboy music, its lyric content often features sad life episodes and generally romantic or melancholic. Country is often parodied by a lone trucker lamenting the fact that his wife has run off while he was away, and even worse, she took his dog with her! Many sub genres now exist: Country & Western, Country Gospel, etc. Contemporary Country is the format most commonly heard.

DANCE While dance music has existed for a hundred years or more, the recent use of the term in radio programming refers to dance floor hits of the current generation. The term has several sub-genres, such as CHR-Rhythmic and the slightly older-skewed Rhythmic AC.

DISCO The term began in France in the early 60s as Discotheque, simply a place where people listened to records. By the late sixties small venues, with a DJ began calling themselves discotheques. In the early 1970's the word was shortened to disco. In the USA this term became was used around the mid-1970's for predominantly soul-pop music typified by the Gamble-Huff music from Philadelphia. The term was later adopted by many pop acts, such as the Bee Gees in the *Saturday Night Fever* era.

DOO-WOP This genre was developed in Afro-American communities in large American cities and is based on vocal harmonies. It soon crossed over with many black vocal harmony groups such as the Drifters making the big time after beginning as doo-wop artistes. As well as vocal harmonies the genre is known for nonsense syllables, a simple beat, very little or zero instrumentation (now called 'a cappella'). Often Doo-Wop uses vocal harmonies to substitute for musical instruments.

EASY LISTENING is often instrument led, with gentle vocals, easy on the ear. Very popular as background music and often referred to as 'elevator muzak', usually as a derogatory term. Quite often this comprises

ONLINE Radio

anonymous and innocuous cover versions of hit songs. In Europe the term 'easy' is often termed 'Middle of the Road' (or, when its poorly focussed, 'All over the Road'!)

ELECTRONIC Stations formatting electronic are typified by a hypnotic bass line and all manner of synthesizer sounds. Electronic music is often heard as Dance music. Today's leading proponents include many who began as disc jockeys, artistes like Tiesto, Daft Punk and Fatboy Slim. Sub-genres of electronic music include Acid-House, Breakbeat, Jungle, Trance, Techno and Trip Hop.

ETHNIC A term often used interchangeably with World Music, it has come to be used for music of one particular nation or region. Some World Music stations feature mainly music from African performers.

FILMTRACKS Soundtrack music from movies were common place in the late 20th century and featured mainly instrumental soundtrack music from well-known films or movies. The owners of recordings are happy to waive copyright payments to encourage radio stations to help promote the film and some commercial stations have playlisted film soundtracks as part of other formats. There was even a film music formatted channel on DAB in Surrey for a while and a hundred online stations follow the format. A good example is the *1.fm* channel from Switzerland

FOLK Quite simply this is 'the music of the people.' At its most basic, folk is cultural, from the roots and backgrounds of people from around the world. Artists such as Bob Dylan, Donovan, Joan Baez, Pete Seeger and Cat Stevens fuse their own distinctive music with more traditional styles of yesterday. The genre includes small niche sub-genres based on the different variations of folk heard in various territories as well as acoustic, Celtic, alternative and contemporary folk.

FREEFORM This could be called formless, it's where the DJ is allowed to play any kind of music with no fixed format to adhere to. It can include any kind of music, even in the same time period. The Freeform format is common in BBC local radio stations and also markets where formatted radio has not become very competitive.

GLAM A form of rock, Glam evolved from a few British artistes who dressed with overt style and glamour in the early 70s to perform pop-rock. David Bowie was the proponent and best known and he has proved to be one of the longest lasting, but others associated with glam rock included Marc Bolan, T Rex, The Sweet, Elton John, The New York Dolls, Roxy Music,

ONLINE Radio

Mott the Hoople, Mud and Slade. These artists paved the way for Punk and Hair Metal. Sub genres of Glam include Garage, Psychedelic, Rockabilly, Surf and the founders of the genre, British Invasion. The Glamrock Junk Shop epitomised the movement with their 'Bopping with the Ballrooms of Mars' and 'Bubblegum Rock' programmes and there are also several online radio programmes on *Luxuriamusic.com* that fit the genre.

GOSPEL Christian influenced music, often up-tempo and with a large choir either leading or backing the vocals. Gospel is a traditional music genre as well as a radio format. Among the many sub-genres of gospel radio format are Southern Gospel, Contemporary, Adult Black Gospel, traditional, inspirational, contemporary hit gospel and rock gospel. One of the best and most easily accessible examples of a gospel radio station is 'Mexican Soul Food', available terrestrially and online.

HARD ROCK This is the format for rock fans who like to feel their music loud, somewhere up past mark 11 on the volume control! From Black Sabbath to Metallica; Led Zeppelin to Foo Fighters; Van Halen to Tool. Sizzling guitars and thumping drums, often augmented by powerful blasts on Korg keyboards. There are at least a few hundred hard rock stations online and at least two on the FM dial. Hard Rock has spun off its own sub-genres; Jam, Prog / Art Rock and Psychedelic to name but a few.

HEAVY METAL A very popular sub-genre of Metal, (see below) Heavy Metal now has around 60 online radio stations and two FM outlets playing the sub-genre. Heavy Metal has its origins in blues and includes artistes such as Led Zeppelin, Iron Butterfly, Blue Cheer, Black Sabbath and Deep Purple.

HIP-HOP Hip hop music is often called rap music. It is a music genre of very stylised rhythmic music over which artistes usually chant. Hip-Hop is a sub-culture which incorporates beatboxing (making instrumental sounds with the mouth into a microphone) and MCing as well as occasionally graffiti painting while doing so!

HOLIDAY An alternative name for the format of playing predominantly Christmas music, usually as a temporary format aimed at shop or restaurant operators to run up to the Christmas holidays.

HOT ADULT AC There are now several different versions of Hot AC in use. It originally meant stations playing current hit music that is more energetic or upbeat than the average Adult Contemporary station, but over the last five years has increasingly meant stations that play a larger proportion of

ONLINE Radio

the very latest releases. Billboard's Adult Top 40 chart includes new releases played on rock oriented CHR stations as well as Hot AC stations.

HOUSE A sub-genre of electronic dance music that has heavy bass chords and its origins in Chicago. The sound has repetitive 4/4 beats, lots of off-beat hi-hats and urgent pounding basslines. The style was popularised by MARRS, S Express, Grandmaster Flash and Sylvester, who each had huge hits in many countries with the genre. Its popularity continues with the likes of Robin Schulz and Duke Dumont and there are several dozen house music-based radio stations around the world.

INSPIRATIONAL The music genre 'inspirational' is generally slow, soft, peaceful and usually instrumental music. Calm, melodic and designed to relax its listeners and inspire fresh thoughts. There are many examples to be found on Soundcloud.

The radio format known as Inspirational contains elements of Blues, Country and Jazz, and is designed to uplift and motivate. Music that makes you 'feel good' and inspires! It is a sub-genre of Christian radio programming that is increasingly popular in the USA. In the UK, UCB run a channel called UCB2 - Inspirational Radio. Live365 have almost 300 Inspirational radio stations on their system.

INSTORE Many large chains of retail stories have in-house radio channels which are used to promote special offers and provide other announcements. Stations generally play well known tunes designed to uplift shoppers and get them in the mood for buying. They are also used for staff training outside hours. *Asda FM* reaches 18 million shoppers and over 150,000 staff every week, making it the biggest radio station by audience size in the UK. The programme service for instore stations is often provided to the store groups by programme contractors. One of the main providers are Imagesound.

www.imagesound.com

ONLINE Radio

INTERNATIONAL A "catch all' which includes radio stations which broadcast 'internationally' such as *RT*, the *BBC World Service* or the *Voice of America* and other stations which play a mixture of international music. Some stations which broadcast a regional type of music such as Hindi, Celtic, Caribbean, Tamil, Middle Eastern and SoCa (from South Caribbean) music are also often called 'international'.

JAMMIN' OLDIES This format is a fusion of predominantly upbeat dance, pop and disco hits from the 60s and 70s, often replacing R&B and Soul stations in some American cities, who are trying to drop older demographics from their audience and refresh their sound with newer and hipper tracks.

JAZZ There are several forms of Jazz and most have been used for radio station formats - around 300 are still broadcasting online and on FM. The main Jazz sub-genres are Acid jazz, Avant Garde, Big Band, Bop, Classic, Cool, Hard Bop, Latin, Smooth, Swing, Vocal and World Fusion.

LATIN This genre of music describes the rhythm and soul of the entire Latin American region. There are many styles, each with its own particular sound. Many contemporary artists such as Gloria Estefan have broadened the appeal of Latin music, which has many different sub-genres; Bachata, Cumbia, Bossa Nova, Banda, Mariachi, Salsa, Merengue, Tejano and Tropicana are just a few!

LITE AC Stations following a Lite AC format include mainly light Adult-oriented Contemporary tracks (singles and album tracks) usually play a high proportion of "easy listening' tracks from the last couple of decades. Not particularly upbeat material, just gentle laid back 'pop' that's easy on the ear.

LOUNGE This is music designed to be heard faintly, i.e. in the background or as mood setting, often Light Jazz, Smooth Jazz or NAC (see below).

METAL "Loud and very distorted electric guitar" is the best way to describe Metal music. Initially made popular by artists like Black Sabbath and Deep Purple, the genre now includes the output of Slayer, Metallica, and Slipknot. Metal has been spun off into a few subgenres such as Industrial, Heavy and Extreme. There is even Rap Metal too and nine internet stations major on just that as a format.

MIDDLE OF THE ROAD has become a little used term these days but until the 1970s it meant any laid-back popular music, often instrumental. See *Easy Listening*.

ONLINE Radio

MOD is a genre which includes mostly up-tempo American soul music popular in the mid 1960s, plus music of the bands who also played those numbers. The Who, Geno Washington and 1970s hits by The JAM are all 'Mod' which also described a sub-culture lifestyle of those who rode scooters and dressed in a 'clean cut' neo-Italian style and were avowed enemies of 'Rockers' who generally wore leather and rode motorcycles.

MODERN ROCK This genre covers mostly current rock music performed by artists who have emerged during the past five to ten years, rather than older established artistes. Most Modern rock is between Rock and Alternative Rock.

NAC This is a relatively new format, more fully *New Adult Contemporary* which has mainly smooth jazz tracks. Most stations are designed to be played more as background music than as a performance, to invite criticism or for careful listening. A modern-day successor to Easy Listening.

NEW AGE Broadly similar to NAC, New Age has become a popular choice for relaxation, inspiration, and enlightenment. Perfect for those who have chosen the lifestyle of yoga, meditation, and other forms of relaxation. Yanni and John Tesh are two of the acts typical of the New Age genre which also has a couple of sub-genres in Environmental, Fusion, Healing, Meditational and Spiritual. Also known as Ambient or Chill, New Age now has over a hundred radio stations using the format online.

NEWS There are very few 'all news' stations, as many former 'all news' stations have now gone over to News Talk, which combines a regular news service with a phone-in format. All news is very expensive to produce as it requires expensive content and only the largest groups or state broadcasters can afford to operate this.

OLDIES This is a very wide term covers music beyond the recent decade or so covered by 'Contemporary', i.e., older than ten years. It has several sub-genres, often 60s, 70s and 80s and even the 50s. Stations filtering music to be simply from a particular era will often sound too varied to attract many listeners, very few of whom would think of their tastes are represented by a particular group of years.
PERSONALITY Stations which present well known or high-profile DJs with extrovert personalities, who connect or engage with the listeners. They will mix interviews, comedy, music and mini productions to form a programme

ONLINE Radio

unique to them. Often, these hosts are so famous that they are given a completely free hand in music selection and it can work well.

Kenny Everett was an excellent example of a British personality DJ. Previously, *BBC Radio 1* had personality DJs and today, *BBC Radio 2* uses mainly famous personalities as presenters, though most useless as DJs, having been indoctrined into other fields, such as television. Key personalities can attract huge publicity and audiences for a radio station, Chris Evans (now back at *Virgin Radio*) being a prime example.

POP Usually lightweight and popular (from where it gets its name) music that is not purely another genre such as country or soul. The term was used in the 50s and 60s for a while and then dropped by music snobs, who profess to not care much for music that was overtly happy or even popular. Often used more as a derogatory term.

P & W 'Praise and Worship' is a sub-genre of Christian programming. It features mainly hymns and gospel songs, 'Praising the Lord' and is very popular throughout many areas of the USA. P&W stations attract a lot of sponsorship and listener funding, scoring good reach.

RAP The American term 'Hip-Hop' never became very popular in the UK. It stands for 'Rhythm and Poetry' and is the chanting of random words and phrases over a rhythmic backing track, something that was used by many black music radio station DJs as far back as the 50s. 'Emperor Rosko' first used it in the UK in 1966 on Radio Caroline and later on Radio 1. In the 80s it became 'hip' for some performers to adopt rap as a music style, backed by laid-back RnB music.

R&B originally known by its full name of Rhythm and Blues, it's based on the BLUES but with a rhythmic beat. R&B became popularised in the late 50s and early 60s in clubs and 'juke joints' (often just a bar with a juke box) by artists such as Chuck Berry, John lee hooker and others who made the slow shuffling Blues songs more suitable for dancing by adding a boogie woogie backing, usually with a solo electric guitar. In the UK, Pye Records added the name R&B to its 'Pye International' label that included such artistes, and many white British groups adopted this as a description of their own style, including the Rolling Stones. In the 90s, some black artistes in the UK adopted the term R&B to describe RAP music.
REGGAE Generally describing the Jamaican music of the late 50s based on calypso and mento, fused with jazz and some rhythm and blues. Some elements of the style were adopted by artists produced by Chris Blackwell in the 60s, notably the Spencer Davis Group and the term became more

ONLINE Radio

popular for the Ska music artists who began having hits in the UK and other European countries after 1966. By 1969, the term Ska had been dropped (although only temporarily) in favour of reggae, which was usually played a little slower than Ska.

ROCK Simply a harder edged pop music, often guitar-led with prominent rhythmic and usually played at a high volume. Perhaps one of the most popular format in all but the most tightly regulated radio markets, rock has shaped the international music market since the 1960s, although the genre itself has gone through an evolutionary process. There are at least a thousand hard rock stations online and on the FM dial.

ROCK AC The AC part of the Rock AC format stands for Adult Contemporary, so it's Rock music (as defined above) from the previous fifteen years or so with more appeal to adults than teenagers. Often with a heavy skew towards adult males, it's a format very popular with advertisers.

ROCK'N'ROLL The rebel yell of a generation of baby boomers. This is at its most basic the integration of black and white music which in the mid-fifties had both parents and preachers absolutely appalled. Upbeat with a fast tempo, Rock'n'roll was often, but not exclusively, white musicians playing up-beat blues music. Popularised by Elvis Presley's "That's Alright Mama" and "Blue Suede Shoes", Rock'n'roll was invariably easy to dance to and up-tempo. Many musicians argue that you can't have Rock'n'roll without a piano and a bass.

SEASONAL Stations playing non-stop or predominantly Christmas music, usually a temporary format aimed at shop or restaurant operators to run in the appropriate period.

SKA This is a music genre that began in Jamaica in the 50s and was a precursor to reggae. It is widely attributed to legendary producer Byron Lee and seems to come from scat, a style of singing. It's certainly based on off-beat guitar chopping. Millie Small's 1964 hit "My Boy Lollipop" is the biggest selling ska record ever, 7 million copies. The name was revived in the mid 70s in the British Midlands, particularly by artistes on the 2-Tone record label, who mixed the rhythms of Jamaican music with the faster paced hard-edged music of the punk era. A typical example of this second phase would be Selector's hit single "On My Radio".

SMOOTH Popular, easy-listening music with a calming effect and a laid back style of presentation, designed to relax listeners and be played in the background. Some large radio groups called it 'Beautiful Music' for a while.

ONLINE Radio

SPORTS As you would expect, this is a mainly all-talk format (although in the UK Rock'n'roll Football, which began as a single programme on *Virgin Radio*, has become a 24/7 format). Sports stations in the UK usually cover many different sports, such as 'Five Sports Extra' in the UK which has grown out of 'Five Live'. It initially offered just additional coverage of lengthy events, such as cricket or race festivals, but is now evolving into a full-time sports only channel.

SOUNDTRACKS Mainly instrumental music is played on soundtrack stations and they are often variations of Easy Listening formats such as NAC or Smooth Jazz. Many of these stations major on film soundtrack music, with lush orchestral backings.

STANDARDS This format is primarily aimed at seniors, older adults (50+). It is sometimes referred to as an 'Adult Standards' or 'Nostalgia' format and is generally full of well -known songs from 30 to 60 years ago done in a non-rock style; i.e., easy listening or MOR ('Middle of the Road', a rarely heard term these days but prevalent in the 60s).

TALK There are many thousands of traditional radio stations which are 'Talk' which simply means all speech, or at least predominantly speech. Most of these are not 'all news' but have lengthy programmes hosted by 'talk jocks' who simply field calls from listeners, or conduct interviews with others. The usual topic of discussion is current events. The main Talk genre has a number of sub genres, such as News Talk, Sports Talk, Christian Talk, etc.

TIS *Traffic Information System* radio is very common in the USA and in some European countries such as Germany. It has been tried in the UK, with one single station previously covering both London and Heathrow Airports, and another covering the motorways to the channel ports in Kent. More recently the Highways Agency ran six different Traffic Radio stations online and via DAB but these has now closed.

TRIPPY Barely a dozen stations claim to format 'Trippy' music; it could also be called Psychedelic Stoned Rock. Some really heavy grooves that are enjoyed by listeners from California to the Netherlands. Other similar stations are *Psychedelic FM, Happy Smokers Trax and Psycho Chill*. The message is in the name and in the music, of course!

TURKISH There are three main strands of Turkish music used for radio station formats; Arabesque, Turkish Folk and Turkish Pop music. Turkish

ONLINE Radio

Music includes many very diverse elements ranging from Central Asian folk music to influences from Byzantine music, Greek music, Ottoman music, Persian music, Balkan music, as well as some styles reflecting the influence of modern European and American genres. There are over seventy Turkish music stations available online.

UNDERGROUND *Air Atlantic* is just one of the 50 or so stations designed to fill the spaces between your ears with some of the most underground vibrations such as Dream, Bliss, Trip Hop, Ambient, Electronica, Indie, Rock & other flights of fancy. *Psychedelic FM* also is an underground station, available only online with its esoteric mix of garage mod pop and underground sounds from 60s to today from around the world. Most underground stations play only album cuts, often quite lengthy ones and would never play a song that is well known or has been a hit.

URBAN These stations tend to be centred in big cities, although not necessarily. They base the majority of their output on popular black music. There are many sub-genres such as Urban Contemporary, Urban Progressive and so on. Urban stations are particularly noted for their crossover appeal making 'black' formats attractive to listeners of all races and so advertisers. The term 'urban' seems to have been coined by Frankie Crocker in the 90s. Crocker said, "A reporter asked what my format was and I told him, quite off the cuff, 'It's what's happening in the city,'; in other words 'Urban Contemporary'"

VARIETY Stations following a Variety format simply use material from a variety of genres. The name is usually used to describe many small non-commercial radio stations in the USA, such as those in the education sector, and some small 'Mom and Pop' radio stations. Such stations often target more mature tastes and many Variety stations will include tracks from various 'classic' genres. There is even a *VarietyIndieRadio* online station in the UK.

ONLINE Radio

Some of the above descriptions of radio station formats in this chapter have been extracted from the book

RADIO FORMATS in the UK and USA.

The book explores the history of how radio developed from being. A patchwork quilt of programming to the many different formats heard today on radio stations. Copies of the full book can be obtained from Amazon:

https://amzn.to/492zgGS

NAME FORMATS There are many formats that are assembled by consultants and then sold by consultants. These are designed to make radio stations sound different to competitors and used in very competitive markets, such as New York or London. Increasingly frustrated with the increasing blurring of the edges of formats, some began giving their formats names, such as **JILL, JACK,** or **BOB,** but these defy definition as they vary so much between cities used.

The Jack FM branding is licensed by an American company called *Sparknet Communications*. It is now heard on four dozen stations in the United States, the UK, Russia and Canada. Its format is more of a station philosophy than a clearly defined musical genre.

Plain old *JackFM* was recently been expanded into a **Jack 2** format, which and became the latest incarnation of former student station *Fusion / Passion / Glide / Glee* etc, in Oxford. In 2023, Jack succumbed the relentless homogenisation of British radio when it was replaced by GHR.

One of Jack's last developments was to respond to listeners by texting them when their selection was about to be played. While SMS messages cost money they are seen by the listener's contacts. This ploy could be useful to online radio stations in growing audience.

ONLINE Radio

Chapter 7

FINANCE

It is important that the founders of an Online Radio station take a long hard look at the prospects for their station and decide as early as possible whether they think it could be a profitable business. Economic viability will depend on the route taken, either on a commercial basis or as a "non-profit making" operation, the majority of which are 'community radio' stations. If it is to be a community station, then the majority of these are geographically small, of appeal only to the residents of a small area.

There are however community radio stations which have very wide appeal. Radio Caroline has identified a market in being a "community of interest' station, enabling it to operate as a community operation, that's of appeal to the widely-scattered demographic of listeners who have a fondness of non-hit album tracks.

Once the correct route is chosen, it's time to consider funding the station. There are many ways to fund a radio station; most fall into one of the categories below. The main stages of a radio station's funding are:

Capital   Licensing, Equipment, Hosting, etc.

Launch costs   Mainly promotion aimed at listeners

Recurring costs   Ongoing operational costs.

FUNDING SOURCES

ONLINE Radio

There are many sources of funding which should all be explored carefully. Each will apply to different station operators in different amounts. Some may not be possible for you.

Capital – Investment from owners own savings.

Syndication – perhaps shared among friends and colleagues.

Borrowing - Raising the money from banks and institutions. By having a formal structured and well-researched Business Plan, an Online Radio station is an investible business, provided that the operators are believable and attractive to investors.

Grants – Many public and charitable bodies support radio stations, particularly those which provide community services or training.

Listener Support is found in many forms. Radio Caroline was an online only station for several years and has had several rounds of fund-rasiing from its listeners; it raised over £100k in the last year that's been ear-marked to renovate their radio ship. Lisbon's *Rádio Quântica* launched a *GoFundMe* campaign in 2023 to stay afloat and cover staff salaries. London online Radio *Balamii* introduced a monthly subscription to cover costs and support emerging artists. These funding options seem to work best with alternative subcultures and communities of interest, such as for fans of classical music and promoting artistes ignored by the established media.

Sponsorship - Getting commercial companies or individuals to make small payments to you, perhaps in return for advertising time on the station. The UK and other radio authorities had strict rules governing sponsorship, however these do not apply to Internet Radio. Provided the content is "legal, decent, honest and truthful."

Radio sponsorship in the UK has always been limited to sponsorship of particular programme items, such as the weather, travel reports or other programme segments. Internet radio stations however are allowed to have entire programmes sponsored and hand over the programme content solely to the advertiser, or sponsor, to make editorial decisions on. They would ordinarily contract out the making of their programme, possibly to an independent programme provider. All the radio station needs to do is load the programme into its audio library and set it to play in its entirety at the agreed time.
Some radio stations carry programmes of an overt religious nature, or even political in content. Internet radio stations have no obligation to be balanced.

ONLINE Radio

There are over a thousand regular AM and FM radio stations in the USA, and many other countries too, which carry 'sponsored programming', usually of 15 or 30 minutes in length, from various religious denominations. The religious ones almost always finish the last few minutes of their time slot soliciting listeners to send them money to continue the Lord's work.

This system has been operating for over seventy years and shows no sign of abating. The preachers of the air clearly think it is good business and will pay to get airtime anywhere they think listeners will support them financially. One of the biggest companies providing the programmes is the *Christian Broadcast Network* who turn over more than $400m every year.

Airtime Sales
The sale of the station's airtime to advertisers. The basic version is as 'spot rates', where an advertiser's message is played as a 'spot', usually of 30 seconds, though 60 second spots are also increasingly common. Spots are often bundled together in groups of four, five or even more, which are known as a 'commercial break'.

For fair and responsible commercial spot sales, there needs to be a system of metrics in place, to ascertain and illustrate exactly who is listening and when. In traditional radio this was done by face-to-face interviews, then by listeners completing written diaries and, more recently, using electronic monitoring. It needs the weight of large numbers to make this system of wearing a tag that 'listens to' audio accurate. Fortunately, Online Radio already has the infrastructure in place where the hosting software can report actual numbers of online receivers are listening to which station as well as for how long and that data can be analysed. This is one of the huge advantages that Online Radio stations have over their competitors.

Listener numbers to online radio are growing but not yet sufficiently impressive to attract large media advertisers to book time on all the hundred thousand or so stations on their air. The onus is on stations to prove to advertisers that their audience are buyers and can generate revenue for the advertiser. Affiliate arrangements are an excellent route and can generate substantial amounts of revenue. A good broadcast consultant can explain in detail and help stations decide which are the most appropriate to their own circumstances as it is a rapidly changing market.

Affiliate sales are discussed overleaf.

ONLINE Radio

Display Advertising
When listeners trust a radio station, especially programme hosts, then they will listen more closely. Persuading listeners to look at a radio station's web site and keep them coming back to it regularly, is best achieved by having information regularly updated and relevant to their interests. This builds up a considerable following, making the web site also attractive to advertisers, who will buy 'display space' on it.

Affiliate Sales
This is simply a commission an 'influencer' earns when leading the listeners to a product which they then buy direct from the advertiser, who passes a commission on each sale back to the station. Listeners and their buying power are valuable. Companies want to sell their products to your listeners, but proving the value can be difficult. Demonstrating effectiveness by the new sales generated by commercials on any particular radio station, is irrefutable evidence of the efficacy of a station's airtime.

These need not be large banner display adverts, but can be simple text links to information pages. Text links such as those you find in this book which link to various information on other people's web sites which are an helpful service to users. The link hidden behind a simple link can contain complex but hidden instructions which computers can read and act on it in a fraction of a second. Typically, a link might say "Here is a potential customer sent over by *Radio Bridlington*; if they buy something from you please pay any appropriate commission to RB". Most of the biggest suppliers in the world, including almost all the main 'High Street' stores and particularly those who sell online, are already set up to do this and they will automatically give a small slice of the value of any subsequent sale to you as an 'affiliate commission'.

Major retailers might offer only a couple of per cent commission but over a period this can add up. Stations need someone to work at keeping the web site updated but it can generate considerable volumes of business.

LINKS TO ADVERTISERS
The links are best if included in a radio station's web site which not only ensures that the lengthy and complex html links are faithfully added to the listener's query to the advertiser, but they are much 'neater' done inside a click. Long URL links should never be read out "on the air". They look and sound ugly and give adverts a cluttered sound. Driving listeners to one's website is not just common sense, it's essential to maximise revenue.

ONLINE Radio

Recurring Costs
These are the costs to sustain the radio station once operational. The main items of expenditure to be provided for are:

Premises
Can be an existing property, even your home. New promises for a station need a lease that should be checked by a lawyer. Onerous terms that might render your station unviable.

Staff
Will staff be employed or act as independent self-employed. It's vital to get this aspect of your operation right, not only for good staff relationships but also from a legal point of view.

Power
Light and heat too, as staff can't work properly in a cold environment and there are strict H&S rules about heat in premises where people work. Air conditioning will also be needed for the optimum performance of both equipment and staff.

Server Hosting
It's vital to have this done well; a professional hosting company is best. Expect a fixed amount each month depending on the amount of capacity you need. There are hundreds of ISPs that provide hosting but one should be chosen that is experienced in streaming and offers additional services.

Promotion
This is the process by which you attract and grow your audience. It is the process of building your product, which is your audience; the millions of pairs of ears listening to your station.

Marketing
Once you have an audience you will want to sell time on your radio station to advertisers – this is called marketing. It is the 'selling' of the air time, or rather the renting out of the thousands of pairs of ears, to advertisers.

Advice
The best consultants don't work for nothing, and can SAVE you money, so to make a station truly successful, work with a really good consultancy, such as Worldwide Broadcast Consultants:

https://worldofradio.co.uk/WBC.html

Financial Prospects

Entrepreneurs will have already spotted that there is considerable potential in operating an Online Radio station successfully and garner considerable profits. The question that we are most often asked now is to quantify these. To do so takes considerable skill and research – but a crystal ball is not necessary.

While broadcasting is part creative, it's also based on science and the discipline of the operator following basic business procedures. A lot depends on how the radio station is created, launched and operated. The very conceptual 'design' of the business is of vital importance as is the calibre of the operational team, much of which stems from their training and their attitude.

The cost of equipment has fallen considerably thanks to digitisation and the capital costs of launching a new Online Radio station need not be punitive at all, provided this is done with knowledge. Similarly the other costs, and in particular, the team. Establish 'over the air' radio has done itself no favours by losing some of its best team members in recent years. Many are keen to return to the air and can be retained by new stations, provided one knows where to look and how to engage those presenters and participants.

Many traditional stations (BBC and the large commercial networks) have left audiences feeling ignored and seeking a replacement, perhaps a station more in tune with their wishes and desires. This opens the doors for new stations to seize the opportunity and provide a suitable service at perhaps a fraction the cost and using some of the same voices and personalities.

The financial possibilities are appealing. Few people ever got rich in broadcasting overnight, but the basic business plan for radio remains the same; it needs skill and expertise. Guidance from the right quarter should be the first step for anyone seeking handsome profits and a good Return on Investment.

Chapter 8
OPERATIONS

There are many items falling in to the area of Operations and not all of them will apply to every online radio station, but some area so important to a station's success that's its appropriate to quickly mention them.

The first item is **Programming.** This will be the main *reason d'etre* of every radio station, if not that of the operator of the station then certainly of the listener.

This is what motivates them, either to broadcast, or to listen. Perhaps both the listeners and broadcasters share an ideal? This is probably the key item that puts Online Radio, or indeed most independent stations, apart from the large corporates on the radio dial today, whether they are the state and publicly owned system (the BBC) or the few commercial radio giants.

ONLINE Radio

It's important that anyone setting up any station, and especially one in the online sector, consider very carefully the question WHY they are setting up the station. If it's purely a commercial venture and designed to make a profit, then consider HOW the revenues will accrue. It's not enough to simply bang out any old programming and hope that advertisers will bat a path to the station's door hoping to buy a slither of airtime to promote their business.

RESEARCH is the key to the main questions that need to be answered and many key questions need to be dealt with well before other important decisions are made:
1. Is there a market for the chosen programme?
2. Who are the potential listeners?
3. Will anyone want to advertise to those potential listeners?

Once those primary questions are answered, the station operator must consider how to reach two core groups - the listeners and the advertisers.

First however, the radio station needs a TEAM of staff capable of carrying out the work:

    A. **Programme Production**. Must be capable of generating programmes that are listenable and will encourage long listening hours by a large number of people. Creative radio personnel should NEVER be hired on a basis of nepotism, friendship, sex (it happens – quite often!) or how cheap they are to get on board. A bad business model is to hire presenters based on how much they will pay to be on the air!

    B. **Promotion** staff, those who spread the word. They handle a promotional campaign which targets the audience, telling them what a wonderful programme, focussed on their audio wishes is available. Promotional activities include where and how they can find the station. Online was a very difficult medium just a short time ago but it's now becoming ever easier as more and more listeners are using it for their everyday entertainment.

    Once the audience has risen to a certain level, it becomes a marketable proposition; large enough to interest others who want to reach that audience. Those are an online radio station's potential advertisers. Note, the word 'potential'. There is a final vital part of a station's operation that must be implemented:

ONLINE Radio

C. **Marketing,** a term that is so often misunderstood, is persuading the potential advertisers to come on board. Often the most daunting part of any commercial operation. Those who engage the prospective advertisers and convince them of the value of a radio station are often worth their weight in gold which is why they are often the best paid.

Most new radio stations cannot really afford to engage the best advertiser sales team who are unlikely to be interested in the small salaries on offer. While a few lazy ones will be happy to take a regular salary they are not sufficiently incentivised to bring in high levels of sales.

The really hungry sales team members (and it's the hungry ones a new station needs) will want paying according to results, which means a commission on every sale. This intensifies the incentive to close sales.

When engaging sales staff, it can be disparaging to find that these can be the least enthusiastic of then radio station's team. While most will range from "very enthusiastic" to "total anoraks", the typical sales force members are unlikely to be interested in radio and could happily sell anything to anyone, including the somewhat nebulous chunk of airtime. Its invisible, and its effect can't always be quantified.

Sales staff will however, become very enthusiastic if the product (the airtime) is potentially very valuable to the prospective advertisers and likely to grow their sales.

D. Technical
In the world of "over the air" radio there is another important factor to consider, probably first and foremost - the engineering aspects of a station is of paramount importance. Unless listeners can actually HEAR a radio station, then it is all a waste of time. Hearing ether radio stations depends on having transmitters working properly with sufficient power and a suitable frequency. It also depends on having listeners equipped with a suitable receiver and, most importantly, knowing how to find a particular radio station among all the metres and megahertz.

ONLINE Radio

In the Online Radio sector, such questions have been made obsolete by the magic of having most listeners being ready-equipped with suitable equipment. The Smartphone and Smart Speakers have all but solved radio's dilemma, although there are still a few important parts at the studio end of the chain that need suitable technical attention. Most kit comes ready-made "in a box" and usually operates better if not constantly tinkered with. A consulting engineer is more than sufficient; just ensure its someone who knows what they are doing and understands the importance of the above factors.

That's a few home truths that had to be said, as there are so many basically good radio stations that have "gone to the wall" because the owners cannot comprehend the real *reason d'etre*, or how the radio business works. It's vital that several important factors are in place and well understood.

Programming   Promotion   Marketing   Engineering

If any of those four core areas are ignored then the basic Business Plan of a radio station will fail and the entire operation is doomed. Only by giving adequate attention and resources to all four areas is an online radio station likely to succeed.

Programming is an area which this book covers to a degree; firstly the overall programme format.

Promotion is the busine ss of telling prospective listeners all about the radio station, or its programmes. One can have the best radio station in the world, but if the audience doesn't know it's there and how to tune in, then the project is pretty much a wasted effort.

The process of harvesting the millions of pairs of ears that the audience hopefully comprises, and packaging them together for sale, or rather rental, is the next vital stage.

Marketing is the selling of those packages of audience to whoever the sponsor might be. Traditionally, these would be spot advertisers or one of the many forms of sponsorship but there are many more potential stream of revenue for an Online Radio station.

Engineering is simply the nets and bolts of getting the signal out there in a state where it can be accessed and enjoyed by the audience

ONLINE Radio

PROGRAMMING

Once the type and structure of the station output is decided; a niche may be helpful, that can set a station aside. It's vital to first ascertain that there is demand for it, among an audience that is marketable.

VOICES Some online radio stations can operate quite well without voices and might run non-stop music of one genre, or might buy in its programmes. Even so, the very least that is needed is some basic continuity between each programme for identifying the station, or trailing forthcoming programmes. Some stations have tried doing this with AI (artificial intelligence) but so far, the delivery has not proven to be sufficiently warm and welcoming to attract listeners in sufficient numbers.

TALENT. The radio personality is still one of the most reliable ways to success. This has been proven over the last century of radio broadcasting, where programme hosts have become key names in their locality and huge national stars when they have been given the freedom to present in their own natural style. First class examples of radio presenters in living memory have been Kenny Everett, Simon Dee and Tony Blackburn.

Many presenters from the BBC and commercial radio found themselves cast aside recently. Most were replaced by networked presenters, usually for no fault of their own, axed by faceless accountants who did not recognise the value of local stars. These personalities are the lynchpins and cornerstones that helped make radio become such an important part of listeners' everyday lives. Many are waiting for a new outlet for their talents.

SCHEDULES

Successful radio stations need a well-defined format as well as a structure and a basic schedule. Listeners like to know what to expect from a station; constant swapping around of presenters and programmes doesn't inspire confidence among listeners.

Programme structure includes features being delivered at specific parts of the clock hour, or at least in defined segments. Research in mature markets proves that these milestones or waypoints re-assure listeners and build audience. It can also be good to break with the pattern occasionally and offer something a little bit different, as 'spice'. "Something completely different" can capture listeners attention and amuse, as a break in routine. This might be repeating a track (but don't do it too often!) if the presenter is so enthused by it, though it needs to sound as though its "off the cuff" and not contrived.

ONLINE Radio

JINGLES

The station 'sound' is very important and every radio station needs a well-defined house style and way of presenting. That doesn't mean that all the presenters should sound alike and nor does any professional advocate using a standard 'voice' at a station for all positing.

The only time to use just one voice for positioning statements is where the station's management are so weak that they cannot (or sometimes will not) rely on their team to do these important announcements properly. Some use an unnatural voice (which simply sounds silly and annoys listeners) to hammer a radio station name into the minds of the listeners. The best voices to use for these IDs are the station's regular presenters.

The ultimate Station Identification is a set of short songs or catchy musical refrains. The most successful and memorable are well-produced, sung and orchestrated station ID jingles. There have been many excellent practitioners in the jingle world over the last fifty years or so, names such as *PAMS, Pepper Tanner, TM Jingles, JAMS* and *Way Audio Creations* with repertoire so familiar to presenters and listeners alike.

While some early commercial stations had musicians extol the virtue of their products in short songs and ditties ("We Are The Ovaltinies" for example), these were very ad hoc and only the biggest merchants could afford memorable ones. In the UK, radio was controlled by the BBC which accepted no commercials. The only opportunity to reach listeners with commercials was via Radio Luxembourg which broadcast into the country form the continent and who pioneered radio ID jingles in Europe.

Jingle production got more organised in the USA in the 1950s when a commercial production company used professional musicians to syndicate short musical radio station IDs. As the USA had thousands of independently-owned radio stations it was huge and attracted many excellent practitioners.

A few productions houses in the UK made some station IDs but mostly stuck to production of commercials for use on ITV and Radio Luxembourg. It was not until the 1970s that a few made station ID jingles for the new ILR stations; one of the leading producers was *Alfasound*, whose founder Steve England had honed his art on Radio Caroline in the early 1970s.

134

ONLINE Radio

One of Steve's colleagues there was jingle expert Norman Barrington, who is currently producing a History of Jingles movie which will amuse, entertain and perhaps excite radio programmers everywhere. He presented a pilot of this 'live on stage' at a radio conference in August 2023 which was very well received. A sample of Norman's huge collection of jingle demos can be seen and heard on his website;   *https://www.normanb.net*

Radio station jingles should be regarded as audio or sonic logos to be applied to the station, like one would do with a company icon, its style or house colours on packaging or the product. They should be even better quality than the music being played as they will be repeated more often, so need to be quite short to avoid burn-out. Professional companies such as MRC will do different mixes to use in transitions between music tracks of different tempo.

Music Radio Creative   *https://bit.ly/3slEwVy*

A radio station's jingles should set it apart from others, so it's vital to NOT use the same package as others in the same market. Most of all, a good radio jingle must be MEMORABLE, and instantly recognisable, helping listeners be sure it's their favourite station when they are ial-hopping.

Jingle content and types
Stations should adopt a strategy to take even more care when selecting jingles as they do over music choice as it is their stations sound. The tactic of selecting the jingles played and where in the running order must match the programme content. They can be very useful in aiding transitions, perhaps from one tempo to a different pace. **Sweepers** are usually a voice with an added sound effect, often used to transition. **Do'nuts** are jingles with vocals at the start, then a big hole in the middle for presenters to talk over, finished with another vocal flourish. An **A' Capella** jingle is one with no musical background, just a sung vocal component. Conversely, a **bed** is an instrumental piece designed for presenters to talk over.

The World of Radio *website* has a page containing many links to jingle producers and other suppliers. It is also a topic on which any good broadcast consultant can advise operators.

https://worldofradio.co.uk/Jingles.html

ONLINE Radio

Voices

The key to success for any radio station and the bridge of communication to the listener that should also provide another essential ingredient – companionship. The world is full of some very lonely people, many of whom rely on radio for friendship. It's a part of radio's success that's ignored by some of the besuited accountants and management trainees who are running (or should that be ruining?) a lot of radio today.

Voices are the most valuable weapon in a radio station's armoury so it's important they sound welcoming and attract listeners. They must also be commanding, to the front and at times strident. These qualities are vital for a radio station to succeed and deliver its best audiences. They are the subject of webinars and a further book, the details of which will be found on the ORB website shortly. *OnlineRadioBook.com*

Station Identification Announcements (IDs) recite a radio station's name and it's important that these are prominent, or even dominant, in order that the listener will remember them. They often have a 'tag line' or positioning statement appended, that should describe the station's values or sound; if delivered correctly and with conviction, it will become a part of the station's identity in the minds of the listener. This is important, especially when they are completing listening diaries.

VOICE OVERS

To have strong impact, many radio stations will use one particular voice for their station IDs and tag lines, often voiced by a professional "voice over artiste, or a voice actor. He or she may spend hours a day in a studio articulating these, along with some even more lucrative commercials.

Voice Over work is an art; getting the impact, emotion and inflection correct is just part of the talks and not every artiste can do it. This is more than simply reading from a script; voice delivery, projection, the placing of accents and many more facets are of supremely important, so its usually better and invariably quicker to get a professional to do a radio station's voice over work. One of the most proficient is **Music Radio Creations,** who have excellent voices from all over the world and excellent production skills too. MRC can handle radio imaging and branding, ComProd and Drops. Check them out at: *https://bit.ly/3slEwVy*

Acoustics

It's important that speech originating in a radio station should be intelligible and comfortable to listen to. There is nothing worse than hearing someone speak in a room full of echo, reflections and other reverberation artefacts. While some of this can be miked out' using a suitable patterned microphone, even the best models can't mask some of the worst rooms.

Broadcast studios usually need the walls and other surfaces acoustically treating to give a nice 'dead' background with very little reverberation. This is particular important if any of your team are to speak live on the air as your studio may need to have some acoustic treatment. This might be achieved, and at substantially reduced cost, by simply having different shaped items on the walls – the shelves of a record collection work very well in this respect.

Carpet on the floor makes a huge difference, and on the ceiling too will help considerably. Better still are special foam acoustic tiles, which are available in a range of colours to give the right ambience in your studio. You can find an assortment of these on Amazon, with varying costs. A good start might be to buy the AFS Home Studio kit which has two dozen high quality Pro-Acoustic foam wedge tiles plus four corner bass traps, made by *Comfortex Acoustics* in Oldham. All the materials meet the latest standards (CRIB 5 Fire Retardency). They are usually fixed to walls etc using spray adhesive, which costs a few pounds a can.

Home studio kit. *https://amzn.to/3QFmwPv*

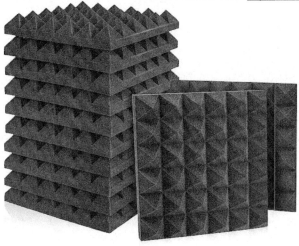

Amazon

ONLINE Radio

Acoustic foam tiles usually cover about one square foot each and can be bought singly. The type of foam used in the AFS kit is specifically formulated for acoustic treatment by a leading British foam manufacturer. It has been in use for many years for sound absorption due to its non-reflective properties. Perfect for acoustic enclosures, studio wall panelling and anywhere else needing sound treatment.

Amazon

Soundproofing mat can also be obtained in convenient 3m long rolls which are 1.25m wide and gives a thickness of about 2mm, which is probably more than adequate, especially if the edges of floorboards and the walls are properly filled first. It's best to use an acoustic sealant for this, which is just £4 a can. Buying a dozen cans gets the price down to just £2 a can, ideal for old buildings or a big multi-studio project.
Soundproofing mat: *https://amzn.to/476Dbkj*   Sealant: *https://amzn.to/45Kavwc*

AC50 sealant has a very strong 'grab' and stick to almost anything. Its often used to stick plasterboard onto walls. Made by Everbuild (one of the leaders in their field) you can paint over it with no problem and its not flammable and is intumescent. Before laying it, make sure that floors are screwed down properly and don't squeak when you walk across them. The mat is also used to soundproof walls; the best results are had when it is between two layers of high-density acoustic plasterboard. That made by Buildershop is excellent, is easy to install and works really well. Details of the Buildershop plasterboard are here: *https://amzn.to/4077GnP*

Acoustic plasterboards are resistant to sound due to the extra weight and heavier make-up of the board. They are designed for use in walls and ceilings where greater levels of sound insulation are required. It's also known as Soundbloc Sound Panel or DBCheck. Blue faced on one side identifies the sheets correctly for orientation. This type of sound resistant plasterboard is available in many sizes.

ONLINE Radio

Acoustic Tile
Amazon

Several makes of acoustic tile are available in the UK and can be obtained from Advanced Acoustics, Acoustic Foam, EQ Acoustics and E-Foam. Most are a dull charcoal grey, but those made by EQ's tiles are blue, a good alternative to having the dark gloomy grey seen in most studios. All tiles can be cut to shape and size using a simple carpet knife. As much of the studio wiring should be run behind the tiles as possible.   *https://amzn.to/45VM92O*

Not only should radio studios be acoustically treated to control various sound artefacts internally, but it's better to have extraneous noise suppressed. Children playing, car doors slamming, traffic noise, including trains passing by, even in the distance, and aircraft overhead. Perhaps machinery from a noisy garage below the studio, or the whine of extraction or other machinery in nearby properties.

The general hub-bub of everyday life is very easy to get used to, and can go unnoticed after a while, but it can sound very bad if picked up by your radio station's microphones. It will soon annoy the listeners and may lead to them tuning out. Bear in mind that such noises may only occur at different times of the day, or on different days.
Always listen very carefully!

Operators should always take some time to sit quietly in any proposed studio location at different times of day and make a note of what is heard. Audible noise should be gauged by calculating the 'noise floor' and the dynamic range needed while using a microphone. It is easy to measure the noise floor and background noise, by using a decibel meter.

Measuring Sound
DeciBel meters were, until very recently, very expensive to acquire, and needed constant calibration. They are now much cheaper and you can even get apps for them to use on a tablet or even a smartphone. Many of the phone apps are surprisingly accurate and one of the best that we have found is also FREE. It's called the dB, the acronym for deciBel, which is the scientific unit of sound intensity, named after American inventor, Alexander Graham Bell. The all-in-one **dB app** is very comprehensive and gives users a dial, plus a graphic and digital representation of average, max and peak sound levels.

ONLINE Radio

ONLINE Radio

Chapter 9

Transmission & Distribution

Transmission is the part of the process of Online Broadcasting in which the servers distribute the programme on to listeners over the internet. The purest form of audio is an analogue signal, sent as a continuous stream of information. If the link is perfect, an analogue signal is always better than the digital one, in quality terms.

Big savings can be made if analogue signals are first digitised before transmission, whether through cables or over radio waves. In the digitisation process, the analogue signal is cut into narrow slices; the average levels are then used, which can be represented numerically. To send the audio programmes uncompressed would take huge amounts of data, Adding some compression reduces the size making it more manageable.

ONLINE Radio

CODECS
A Codec is a device or software that processes the signal and makes is totable for sending over a network. Previously a codec meant a combined coder and decoder but today they usually perform only one function. Most codecs are for playback (decoding) the signals, but the one used to encode outgoing audio is also called a Codec. Choosing the right CODEC is important, as is the right CONTAINER FORMAT.

Own server or hosted?
It's possible to broadcast an online station using any server with internet access. It is preferable to have the server located at a professional server 'farm' with plenty of bandwidth to the listeners. Most small-scale Online Radio stations buy in such a service, usually from their ISP host or a consolidator, as they will have many more potential connections into the main networks and versatility for expansion 'on demand'.

Online Radio Station software
RadioStation Pro is a professional-grade, fully-featured WordPress plug-in for broadcasters of any type, especially online radio stations. It's used to create and maintain a schedule and the metadata on stations' websites, while providing listeners with uninterrupted streaming audio playback.

RadioStation PRO has been designed especially for Online Radio broadcasters. It enables them to produce expert management programme schedules and optimises display to listeners. Content can be integrated using schedules, shows, profiles, episodes, playlists and more, making stations sound more dynamic. There is also a live streaming audio player called *StreamPlayer* PRO which gives listeners a nonstop stream while on the station's website and is now available as an add-on component.

Support for stations using StreamPlayer and Radio Station PRO is looked after by *Netmix* and is excellent. It's available to purchase outright, or can used on either a monthly ($10) or an annual subscription. Stations can try it free for a couple of weeks and there is a 100% no-risk, money-back guarantee. **Radio Station PRO** have made extensive efforts to ensure that it includes all the things a discerning website developer would expect from a plugin, which is why it's the market leader. Their website has a full list of the features available for each variant of the software and links to many tutorials, interviews and some further help on how to improve radio stations by founder Tony Zeoli.

Radio Station PRO
powered by netmix

See this link for information:
https://r.freemius.com/4526/7605466/

ONLINE Radio

AUDIO HIJACK

Audio Hijack Logo
Rogue Amoeba

We recommend Audio Hijack for broadcasting from your Mac. The Audio Hijack application includes nearly all the functionality Nicecast had, along with many additional features and improvements. With the Audio HiJack Broadcast block, one can stream MP3 or AAC audio from a Mac to an internet streaming server powered by Shoutcast or by Icecast. It's perfect for running live feeds of podcasts, as well powering online radio stations and more. Click here to learn more and download the free Audio Hijack trial.

https://www.rogueamoeba.com/audiohijack/

ICECAST

IceCast is a streaming media project released as free software by the Xiph.Org Foundation. It supports AAC+, Ogg Vorbis, OPUS audio and MP3 audio streams and can be used to create an Internet radio station or a privately running jukebox. New formats can be added easily and it supports open standards for communication and interaction.

Each IceCast server can house multiple streams, which are called mountpoints. Stations can have a single Icecast server containing multiple broadcasts, each with different content, or possibly the same broadcast but with streams of different bitrates or qualities, used to send low bit rate versions of a programme. This helps avoid listeners paying high rates for data consumption. There's little point using data if the quality resulting is simply wasted on a mobile phone or in a car.

BUTT

An odd name for a piece of important and FREE software app (**B**roadcast **U**sing **T**his **T**ool) that runs on Mac, Linux or Windows. It's an easy to use, multi-OS streaming tool and can handle SHOUTcast and Icecast. The main purpose of BUTT is to stream live audio data from a computer onto a Shoutcast or Icecast server. While recording is also possible, BUTT is NOT intended to act as a server itself or automatically stream audio files. Supports MP3 and OGG files, reconnects on 'auto-start' and has a status display. The download and info files are on this website:

http://danielnoethen.de

ONLINE Radio

Listen2MyRadio. *https://rb.gy/chobt*

Established in 2006 as the first free Shoutcast hosting company on the internet, *Listen2MyRadio* remains the biggest free Shoutcast hosting company. More than a million users have signed up to their free and premium services. They offer all aspects of streaming over the internet from servers in the UK, USA, Canada and Germany. There is a free access service, supported by adverts, or a monthly charge which accommodates up to 5,000 listeners. There is no limit on bandwidth.

The Listen2MyRadio system has a fully featured, web-based media administration and automation control panel utility which allows radio station owners to manage their *Listen2MyRadio* services in a secure environment. For the Premium accounts this is the highly regarded Centova Cast control panel. It is user-friendly, can configure your stream for you automatically, give comprehensive statistics and royalty reports as well as feature rich default station web pages.

Listen2MyRadio uses a variety of technology including *SHOUTcast, Wowza, Windows Media, Virtuozzo* and *Centova*. Stations can stash 5 GB (or more on request) of files in their account for use with autoDJ allowing their station to continue playing even when unmanned. Stations can have their own bespoke *Media Player* added to their web site.

Stations can also be heard on their own mobile apps which plays in the background while other apps are in use. It can give links to 'now playing, radio station web site, contact the studio, and other useful features that make listening a rewarding experience. Instant setup is available for Listen2MyRadio and it takes less than hour to get a station online.

https://rb.gy/chobt

SHOUTCHEAP

Online Radio Hosting services at affordable prices. The ShoutCheap I.T. team provide clients with SHOUTcast and ICEcast streaming technical support or help with launches. As well as offering a 'no contract' arrangements so that stations can cancel at any time, their Freedom Plans enable online stations to customise according to their needs.

Freedom gives an online radio station complete flexibility on deciding whether to use ShoutCast, Icecast, CloudAutoDJ perhaps them all. Shoutcheap's accounts have unlimited traffic but with limits on bitrate and listeners. The options begin as low as 20p a day and all plans include the CloudAutoDJ with all its great features for free.

ONLINE Radio

Broadcast Software for Apple Macs

- **AUDIO HIJACK** A virtual audio router, full of excellent features.
- **MIXXV2.** comes with an inbuilt encoder & can feed multiple servers simultaneously. MixxV2 includes a free Mac encoder.
- **BUTT** 'Broadcast Using this Tool' has a free Mac encoder. It's easy to use, and has multi OS streaming tool that supports SHOUTcast and Icecast and runs on most Operating Systems.
- **RADIOLOGIK** feature-laden software, good for 'live-assist' or fully automated. Has good features for DJs with mixing background.

Broadcast Software for Windows PCs

- **RADIODJ** a free app but not as good as SAM broadcaster
- **PLAYIT LIVE** another free software but with premium plug ins and modules needed which can be expensive.
- **MIXX V2.** One of the best free DJ programmes that's easy to use. Can be complex to use for an MP3 output but has over a dozen sound effects onboard. Screen can become very cluttered.
- **PROPPFREXX ONAIR** An advanced automation system developed in Germany. It's packed with features, is very stable, reliable and versatile as well as working with touch screens. ProppFrexx is good for voice-tracking and newbies. Has two integrated cart walls for jingles and SFX plus many additional features such as the ability to fetch RSS feeds (for news and weather). It can use every format imaginable, including the professional AIFF and WAV ones.
- **VIRTUAL DJ PRO** Been downloaded millions of times and the best known, very simple to use but expensive ($300) for its features.
- **SAM BROADCASTER PRO**. Lots of features and well-respected, Includes many excellent features, all useful running a professional sounding radio programme and it includes a built-in encoder. Very stable, having been around for 22 years now. It's by far the best-selling online radio software. Easy to install and add music and can be operated in three modes.

More details on SAM Broadcaster software: *https://prf.hn/l/xnPLYZz*

There is a more extensive list of other suitable software on the website
OnlineRadioBook.com

ONLINE Radio

Radio.co

The brainchild of James Mulvany, who studied multimedia design at Huddersfield University. *Radio.co* is a full radio streaming system offered by a company located in the centre of Manchester, with a satellite office near San Francisco. It allows anyone to build an online radio station and broadcast globally. Among the early testers and adopters of the system was Sunset Radio, in Manchester. One new entrant to the platform is pop-up station *Radio Everyone*, which is part of writer–director Richard Curtis's Global Goals venture.

The *Radio.co* platform is aimed at a variety of radio operators, from traditional or conventional broadcasters through various media operators down to the smallest 'hobby' radio stations. Since Radio.co launched in Summer 2015 they have built their service to deliver almost a thousand streams and 35 million hours of listening.

Their online radio package includes a full automation system that is capable of live broadcast. Presenters can break into the stream remotely from wherever they are. Radio.co also offers analytics, station management, a customisable web player, a fully optimised stream suitable for reception on mobiles as well as 'easy to operate' integration with most platforms of social media.

The Radio.co web site includes their own 'university' of Internet and Online Radio and streaming which offers full product details, an overview of their dashboard on which online stations can monitor their output statistics plus a Learning Media section which teaches how to upload files, create tags and build playlists. It also demonstrates how to build a player that can promote an online radio station on social media, such as X and Facebook.

Live365

A long-established distributor of Online Radio that provided the tools to programme, earn a share of revenue and submit music returns. Around 2016, they closed and the name was sold, but now appear to be offering their service again. We have found them uncommunicative.

ONLINE Radio

AGGEGATORS DIRECTORIES
(Online station listings).

One of the biggest problems that new radio stations encounter is that there are so many stations available that they can be very difficult to find, or get found. This is particularly problematic unless listeners are already searching for a station, because they have never heard of it. The chance of them stumbling across the URL of an online radio stream is negligible, even if they are seeking new stations to listen to.

Radio directories are simply listings of radio stations, usually grouped by category or by genre. These make it easier for listeners to find stations that are of interest to them. The directories don't 'host' any broadcasts, they simply switch listeners to the originating streams when they click on the link in the directory. They are assembled and operated by 'content aggregators' many of which are described below.

Many directories have a mobile App enabling listeners to tune in using that. Most have versions for all formats of mobile including iPhones, Android and Windows handsets. The directories also help radio stations get referenced on search engines so, it makes sense to get online stations listed on as many of the directories as possible.

Very few directories have a charge for adding a station to their guides.

AUDIOREALM

This is a part of the same company (Triton Digital), who own *Sam Broadcaster* and *Spacial Hosting*. Their directory has just over 1,300 stations; all the stations listed in their directory use Spacial products and services. The radios stations carried on AudioRealm broadcast many formats, from Top 40 to rare niche music genres. By becoming a member, listeners can store a list of favourite stations and AudioRealm suggests new radio stations. *www.Audiorealm.com*

DELICAST.

DeliCast is a directory of streaming media which is freely available. They are not content providers, simply the information about broadcasters, along with corresponding links. Delicast's aim is to make it easier for Internet users to find broadcasters' websites and their streaming content. Access to the Delicast website is free of charge. Listing thousands of stations, both Internet Radio and tv, DeliCast is a simple directory listing by country, genre, popularity.

www.delicast.com

ONLINE Radio

InternetRadio
The directory part of InterRadio has almost 50,000 radio stations linked to it. To add your station to InternetRadio's directory click the link below. As well as the directory they also offer streaming services, with SHOUTcast and Icecast servers in London and in Dallas. InternetRadio uses the *Centova* control panel whose statistics make stream configuration easy.

https://www.internet-radio.com/add/

LiveOnline Radio
LiveOnlineRadio is a directory of music streaming stations which provides online radio and music stations with various genres. Links to external sites are opened in a new window to either the appropriate media player, or to the content site where possible. Stations listed are from all around the world. New stations must complete a form on their website:

https://liveonlineradio.net/submit

MyTuner
A directory of radio stations and podcasts owned by *AppGeneration – Software Technologies Lda*. myTuner Radio has over 50,000 radio stations and one million podcasts from all around the world. New online radio stations can have their output added by completing their form. With a modern and easy to use interface, the myTuner radio app gives you the best listening experience in all your favourite devices anytime, anywhere.

RADIOFEEDS UK & Ireland
A unique and innovative way of listening to the UK and Ireland's local radio stations on the internet! Here on RadioMaps.co.uk, you can tune in to **all** the webcasting local radio stations in the UK and ROI. The RadioFeeds UK & Ireland site, offers more information on the stations featured on the maps.

http://www.radiofeeds.co.uk

RADIOGUIDE.FM
Not an FM guide at all but a directory of online radio stations and a good resource for new stations. Add online radio station to their *radioguide.FM* and create you're a dedicated station page, increasing your exposure worldwide. Online station owners can promote stations on the x portal.

https://www.radioguide.fm

RADIONOMY
Also known as the **Radio Guide FM**, Radionomy is a free platform for online radio, providing broadcasters and listeners with the tools and infrastructure to create, broadcast, promote and monetize their own online radio stations. Submission is by filling out their form: *https://www.radioguide.fm/add-station*.

ONLINE Radio

SCREAMER RADIO
Screamer Radio is a freeware Internet Radio player for listening to radio on the internet. Its development began in 2003. Screamer's player is advert free, has a peak meter and thousands of radio stations. It supports *Shoutcast, Icecast (MP3* and *Ogg/Vorbis)* AAC streaming and WMA.

https://www.screamer-radio.com

SIMPLE RADIO
Released in 2014 by Streema, one of the easiest to use and perhaps most reliable apps for iPhones, offering thousands of stations that can be set as "favourites" and a sleep timer. A Juke Box facility allows selection of a format and offers random selections.

www.simpleradio.com

StreamingThe.net
StreamingThe.Net has over a decade of expertise in online media aggregation. It's easy to add stations to the STN directory which works on most mobiles without an app download being required.

https://www.streamingthe.net

STREEMA
Streema was started by three guys (Richard, JT and Martin) who believed that tuning in to your favourite stations should be as a simple as possible; easy, fast and fun. The Streema app offering more than 70,000 radio stations and watch more than 10,000 TV stations in its directory.

https://streema.com/radios

TUNE IN
A global audio streaming service providing news, radio, sports, music, and podcasts to over 75 million users. The company was founded by Bill Moore in Texas in 2002 and is now based in San Francisco. Listeners can listen to online radio using the TuneIn website and is also available on more than 60 models of car. Tune In is well-funded and has raised more than $47 million in venture capital.

TuneIn has done partnership deals with many other broadcasters around the world, including Amazon, where TuneIn's *premium* service is now available on Alexa devices and with News UK, bringing news, music and sports coverage to the platform. **TuneIn Radio** is a free app with access to 100,000 radio stations and podcasts.

http://www.tunein.com

ONLINE Radio

VTuner
With the vTuner technology integrated into products, consumers can access live radio stations, podcasts and on-demand shows free from around the globe. V-Tuner have over fifteen years of experience at delivering a 'best-in-class' Internet Radio experience. The VTuner platform is used by many ICE brands give easy access to online radio for listeners.

https://www.vtuner.com

VLC Media
The VLC media player is popular, free cross-platform media player developed by the *VideoLAN* project. VLC is available for desktop operating systems and mobile platforms, such as Android, iOS and iPadOS. It supports almost all audio and video formats without the need for additional codecs. Fans of internet radio stations can stream free of charge on the VLC media player using the *Icecast* feature.

YTuner
YCast is a self-hosted replacement for the vTuner internet radio service which many AVRs use. It emulates a vTuner backend to provide an AVR with the necessary information to play self-defined categorized internet radio stations.

Wikipedia
Wiki also has a listing of radio stations with links to them. It is far from complete but provides a good way to let people know of a station. At the end of 2023, the Wikipedia still called it 'internet radio' and showed links to the large groups or country listings, but it's possible to add links to individual online stations.

Windows Media Radio
A component of Microsoft's Windows OS, WMR is now part of their 'Legacy' suffix and has been rebranded as Media Player in the Windows 11 OS. It uses the ASF, WMA and WMV formats. It's also used extensively for ripping audio from CDs. There have been many security issues with WMP. *Windows Media Runtime* in Windows 2000, XP, Vista and Server had a bug that permitted remote code execution when a user opened a specially crafted media file. This file allowed attackers to install, change or delete data and create new accounts on the user's PC with full admin' privileges and rights. To add an online station's details into the *WMR* directory, email *tuner@microsoft.com*.

Chapter 10

ONLINE RADIO RECEIVERS

A History of receivers (from crystal sets to smartphones)
The first radio receivers were all home-made, mostly cobbled together from surplus military equipment, or even hand-made components such as coils and capacitors. Receivers were invariably very simply constructed, using a crystal detector, a coil and headphones. When put together, this was known as a radio 'set'. They were connected to large aerial array of wire which was the antenna, strung out along the garden or, in a suburban home, often around a picture rail.

The crystal detector was often known as a 'cats' whisker' as it was a small wire touching a small piece of a crystalline material such as galena. It rectified and demodulated the incoming radio signal, leaving only the audio modulation, which could drive a small headset.

ONLINE Radio

"Listening in" as it was called needed a pair of high impedance headphones camped onto the head. This was because the signals produced were so weak and one needed to exclude any extraneous noise. The demodulated signal was not powerful enough to drive a loudspeaker. Valves were a real 'game changer' and meant that the whole family (often neighbours too!) could gather around the radio and listen to the far-away voices and sounds; radio really was a technological marvel.

Leonard Plugge & his car aerial
IBC

The first in car radios were developed by Captain Leonard Plugge, a true radio pioneer who bought airtime on foreign stations and broadcast programmes in English with commercials back to the UK. Plugge built the first car radio and erected a loop antenna onto his car which he then used to tour Europe signing up new stations to relay his programmes from London.

Commercial radio was not allowed in the UK, the GPO refused to give a licence for it, so some radio stations beamed programmes into Britain from other European countries.

This was the first example of cross-border radio to circumnavigate the authorities' strange-hold on broadcasting, which prevented newcomers or independent companies getting access to radio. Today's online radio stations are the successors to stations such as those pioneered by Leonard Plugge almost a hundred years ago.

Portable radios began to appear in the early 1930s, using miniature valves but these were power hungry, expensive and the preserve of the rich. The next landmark in radio receivers was a 1947 invention, the transistor. These small devices were semi-conductors that rectified and amplified signals, replacing the bulky and power-hungry valves. The transistor made it possible to have a truly portable radio.

The transistor revolutionised listening as it made radio more personal, freeing listeners from the huge valve set in the home. Teenagers finally had their own device for music and it became a status symbol.

ONLINE Radio

1970s radio-cassette recorder
Amazon

A huge development in the sixties was the 'Compact Cassette' recorder, which was soon joined to a radio to make the 1970's icon, the radio-cassette recorder. Now the listeners were able to record direct from the radio, which invariably meant recording the Top 20 hits countdown shows, when the hits were played one after another. Fingers really worked overtime on the 'pause' button to remove the DJ dialogue!

Compact Cassettes soon replaced a failed piece of consumer technology found mainly in cars, the 8-track cartridge player. These never took off in the UK, due to their cost.

The radio receiver's next development was the integrated circuit, usually called 'chips'. They were simply arrays of transistors on a single slice of semi-conductor (the chip) making it possible to put the various stages of a radio onto one component. This greatly reduced the size of radios and other electronic circuit. A problem with this incessant drive to miniaturisation that ceramic filters were used, causing poorer sound quality of AM signals.

Miniaturisation also meant that thousands and then millions of circuits could be accommodated in domestic equipment and that digital circuitry could be mass produced, enabling huge efficiencies. Digitising electronic signals has enabled digital audio, images (both still photography and video) and data transmission. That in turn has made it possible to transmit it much more efficiently and squeeze millions of channels of audio down single cables, or a glass fibre.

Miniaturisation and cost efficiencies have revolutionised all communications and entertainment, as evidenced by todays smartphones which are capable of doing almost anything. Ignoring the many features of smartphones, even regular radios today have a wealth of features, most of which would have been impossible a few decades ago.

Among the advances made were better audio quality, then pairs of channels enabling stereo sound, just as our ears naturally give us. The quest was then on to expand the capacity of the spectrum for radio to carry more stations.

ONLINE Radio

Digital radio meant that the space given to each channel could be reduced and each programme could 'time share' a single carrier. The big problem was that almost everyone had a radio, or access ton one and the new digital modes required new receivers.

When the BBC began transmitting DAB in the 1990s, manufacturers were reluctant to mass produce the new receivers. Early models were flawed with heavy power consumption and DAB still lacked popular programming. The advent of new stations only available on DAB made it more popular, as did lowering of prices. When the improved DAB+ was launched, even more stations could be accommodated, but the UK broadcasters would not switch for fear of upsetting the few million who had bought the original radios.

By the early 2020s, there were over 140 different models of receivers available most of them with DAB+. Many have other bands too; invariably FM and a few dozen have Medium Wave with half a dozen still on sale with the Long Wave and some have the SW bands too.

Most personal radios are priced at up to £50, although there are some up to £100 each. Most of the table-top receivers and other mains driven radios have both DAB and DAB+ available; they cost from £70 to over £2,300 for the **Ruark** R7. This is a floor standing, complete AV solution that offers listeners Online Radio, DAB and FM with synchronised sound around the home. Ruark are delighted to add new stations to the database of URLs used by their radios. There are several other models in the RUARK range but they don't tune to Online Radio, offering only DAB.   *www.ruarkaudio.com/*

The Ruark R7

The PURE Elan Connect is a useful portable Online Radio offering 25,000 stations with the option of DAB+ and FM too. The 2 ½ inch full colour display makes station selection a breeze, and there are 20 presets for each band, including online stations from the internet. All the PURE *Elan* series receive online radio.   *https://amzn.to/3s9M4ux*

See more **online radio receivers** in the next few pages and the latest details of the models available by mail order.

https://amzn.to/491Xp0g

ONLINE Radio

Online Radio Receivers

Even though online radio has been on the air for almost thirty years, the number of dedicated receivers for it has never grown much with just a few specialist models available, and none at competitive prices for equipment on rival platforms, such as FM then DAB.

Online Radio can be tuned using any computer with a web browser and as that now includes the 6.9 billion smartphones, one has to ask how much demand there is for a stand-alone receiver in its traditional form – a box like structure that simply receives Online Radio broadcasts, or even one that can hear every kind of radio station? The trend today is towards the use of a combined unit in a small hand-held device that can be slipped in the pocket. It includes camera, TV, computer, word-processor, spreadsheet and everything that one could need.

There are now many models of radio which offer both DAB, and DAB+ as well as an online section too, enabling the listener to tune into many tens of thousands of stations from around the world.

One of the cheapest Online Radio receivers is the LEMEGA IR1 (we think it stands for *Internet Radio* 1?). It offers not only clear reception of online stations but also those DAB, DAB+ and FM too. It has a clock with dual alarms and sleep / snooze facilities and Blue Tooth.

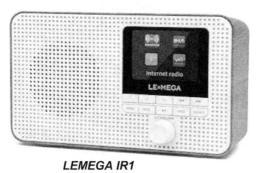

LEMEGA IR1
Amazon

With 60 presets and a headphone output socket, the LEMEGA is very versatile and all functions are controlled from the front by push-buttons, with just a solitary rotary off / volume control, making it a good simple radio for the home. Powered by mains or by a couple of AA cells, the IR1 is now on sale for about £55.   *(https://amzn.to/46uE6Lb)*

ONLINE Radio

Roberts have been a much-revered radio manufacturer for over ninety years and have their HQ at Mexborough in Yorkshire. Their model *Stream 83i* portable radio had FM, DAB and online radio making it very versatile. The audio fidelity was excellent and we happily paid £160 for one several years ago, which is still going strong and has never let us down. Its three speakers give very clear sound and its piano black finish looks good too. Used examples can now be found for around £75 on EBay.

Roberts Online Radio 94L

It was upgraded to the 94i and a 94L with a nice-looking acoustically-tuned wooden cabinet, for just under £200. It offers speedier start-up, a more powerful processor, better audio quality and easier to operate controls as well as Amazon Prime Music and Deezer functions. They cost just under £200 on Amazon.
https://amzn.to/46SiHLL

Roberts also make a **Revival iStream** model (93L) in the 1950s style of their popular 'traditional receivers and its available in several colours.
https://amzn.to/3FpBeDH

Roberts Radio

UK-based Frontier's Majority arm has an Online Radio with BlueTooth, a remote controller and a built-in sub-woofer using a four-inch speaker. It plays many online stations, plus DAB and DAB+. It has 90 presets and a colour screen. These radios cost about £150 each from Amazon complete with a three-year. They include all the power and connections cables, a remote control and operating guide.
https://amzn.to/402sXPl

ONLINE Radio

An ideal Online Radio receiver?
Some very 'high end' Online Radios, such as the Fitzwilliam models by the *Cambridge Majority* group cost almost £1000 each, and some can handle Podcasts too, making them very useful and versatile. There is also a Majority for less than £150 which has Online stations and Spotify.

Cambridge Majority

So far no one has brought to market our own 'ideal online receiver' which would include a programmable recording device. We could certainly use one that can handle multiple channels and one can set the recording times for a few days in advance; something similar to the Sky Q Box. It is most certainly possible, though the commercial viability and may be what's holding manufacturers back. Perhaps there are very few listeners who could use such a machine, although with the rapidly expanding listening audience for online radio, we are sure such a radio will become available, eventually.

Some retail stores stock internet radios but we have found that the sales staff are often lacking in knowledge. This is one of the areas in which the customer is usually better served **ONLINE**, which is surely the natural home for an online radio receiver! Amazon seem to offer the widest range of online radio receivers and their prices are invariably the keenest. They have a couple of dozen internet capable receivers available, most can be found via this link *https://amzn.to/3M51Ygp*

To list each model and give a meaningful review is beyond the scope of this book. The *Radio Listener Guide*, is an annual publication and in its 156 pages, their experts David Harris and Phil Wright review and compare all the new models on the market in a fair and even-handed manner. It also lists all the terrestrial transmitters and comments on the trends in radio broadcasting. It's a publication we recommend.

https://radioguide.co.uk

ONLINE Radio

Smart Speakers
Considerable growth in online radio has come with the explosion in Smart Speakers, which can respond to an audio request by the listener to tune to certain named radio station. New laws going through Parliament in 2024 will make it mandatory for all Smart Speakers to play all stations licensed by OFCOM without charge and without advertising being added.

When the devices first launched there was concern in some quarters that the devices, like Smart TVs, were always listening and that the voices and information became the property of the suppliers, such as Amazon and Google. This concern seems to have waned, presumably the convenience of not having to tune a radio now outweighs the thought that 'brig brother;' may be listening.

Amazon Alexa smart speakers: *https://amzn.to/3Qm9p44*

The most popular types of Smart Speaker is the Amazon Alexa, thanks to its low price and versatility. Models from Amazon, Google and Sonos are strong sellers too, with the high quality end of audio being taken care of by Bose. Many mobile phones will also work in exactly the same way, Apple's SIRI is proficient at finding stations.

Alexa Skills
To maximise listening on smart speakers stations, stations need their own 'Alexa Skills' commands, so listeners need only say "Alexa, play me (Station name)". These can be written by companies such as *Broadcast Radio Ltd*, who can also add a visual logo to which shows on many newer smart speakers and further drive involvement by letting listeners ask Alexa which song is playing. (see Page 172).

Amazon's Echo Show
The new generation has an alarm clock facility and a five inches smart display. This can show a variety of images, including a calendar, photographs, conference calls, TV shows and even the feed from a front door camera. The Echo Show units have very clear audio from an improved speaker, delivering clearer vocals and deeper bass for truly vibrant and a more dynamic audio experience. These became the #1 Best Seller in 2023, possibly thanks to them now having multiple layers of privacy protection. Not only has the Echo Show got a button to switch the camera and microphone off but it has a built-in camera cover and the ability to delete recordings.

Echo Show 3rd Generation
Amazon

https://amzn.to/3sdAMVY

ONLINE Radio

Online Radio in cars

Before the CoVid pandemic, there were almost 41 million cars, vans, trucks and buses on the road in the UK (DVLA road tax figures) and it's estimated that over 40m are equipped with a radio. Many new cars have online radio receivers built into the dashboard as part of the ICE (in-car entertainment). The main options are the Apple's *CarPlay*, the Android *Auto* or the *Mirror-link* systems. They receive online stations through their tuner using an antenna built into the body of the car.

There are ways to these systems to be added to an older car. The Kodiaq system found in Skoda cars, features online radio and SatNav, with head-up displays. Many new cars have a WiFi "hot-spot" built in, so they can relay web signals to car occupants, receiving it from cellular masts via a car-shell antenna. Wi-Fi plans cost under £20 a month but these can be subscribed short term, such as a few days, making them ideal if embarking on a long road trip or holiday, if the car is relatively unused at other times.

A cheaper way to listen to online radio in the car is using a mobile phone, linked to the car's ICE speakers. This can be done with a USB connector, which many cars have today, or by driving the car's audio from the BlueTooth output from your phone. A small *BlueTooth* to FM adaptor costs less than £20, see HERE for some examples. *https://amzn.to/3tqru9I*

Digital Radio in cars

The EU imposed a law called the European Electronic Communications Code that makes it mandatory to install a digital radio receiver in every new car. It applies only to new cars and it does NOT mandate that the radios should be DAB, only that they should be capable of receiving digital signals, which can be ONLINE radio.

SDRs

The term "online radio receiver" is also used for a tuner plugged onto a computer and accessed over the internet for remote accessing. These SDR tuners (Software Defined Radio) usually receive all frequencies from VLF up to UHF which it converts to digital and resends the audio, online. They enable listeners to sample the airwaves at the remote location but are NOT a part of the ONLINE RADIO that's the subject of this book.

ONLINE Radio

APPS.

An abbreviation of 'application', in the Online Radio context app means simply a programme or instruction routine that allows a computer to receive and decode audio from a data stream. Apps are most often encountered as add-on on mobile devices, such as phone or tablets, but also exist for desktop and laptop PCs too. They make complex routines, such as tuning to an Online Radio station and deciphering its audio and lots more, without any technical knowledge.

Most will require internet access to run and are usually downloaded from 'App Stores'. Many of the apps in general use are 'native' apps, are written in either HTML5 or in CSS and use a browser. They are designed to run on one of the major mobile operating systems: Apple (iOS) or Android and might be supplied as Freeware, for a single 'buy out' payment or on a subscription basis, either monthly or annually.

Developing apps needs careful planning as mobile apps have several pressing needs. The processor in mobile devices is often under-powered, and the app designer has to consider the many screen sizes they might need to run on.

Many of the large radio stations and the distribution companies have their own app which is often made available to stations carried on their network. *TuneIn, SlackerRadio, AccuRadio, IHeart Radio* and *My Tuner* are the most commonly found but we prefer **Simple Radio** for ease of use. It tunes swiftly from station to another. Simple Radio offers a reasonable size list of favourite stations, enabling listener to quickly find the one they want to hear next; that's perfect for those who are promiscuous in their radio listening.

Almost every online radio app has a search facility where it can quickly scour through the many tens of thousands of stations that it has pre-loaded. Most apps can download individual episodes of programmes too; this is called "on demand listening", or Podcasting. I wrote a guide to Podcasting last year and details are available on page 185

https://worldofradio.co.uk/Podcasting.html

One of the most used is the **BBC Sounds** app which gives access to the linear streams (or stations) and its on-screen schedules which allow immediate selection and rewinding. The BBS Sounds app was developed with the promise it would eventually carry other stations too but so far only carries BBC channels. It could revolutionise online radio, if fully developed. Apps are usual written by developers, who will assemble the code, and then submit it to the two main distributors of Apps – Apple's Appstore, and

ONLINE Radio

Google Play. It's best to include the radio station's name in the title so it can be found on the two App stores by end-users.

It is possible to write an App without the intervention of a web-designer. **Appypie** have an AI appmaker that needs no coding and offers easy step-by-step instructions.
appypie.com

AppInstiture.com is a 'no-code development platform' that offers a FREE appbuilder that uses AI to complete some of the stages. It has a 'drag and drop' functionality that unifies and simplifies data into a single source.
appinstitute.com

Android users seem to prefer the AppsGeyser' route which provides the Radio Player template and is based on the official VLC client for Android. .
Appsgeyser.com

Apple make it easy to write an app and start broadcasting an Online Radio station on iPhones. Accessible by simply playing a track in Apple Music and tapping 'more' then "Create Station". Even easier is to ask SIRI to "start a station from (genre of music).

Apps are an important distribution platform; they make it possible for listeners to find, listen to and engage with radio stations. Many operators prefer to delegate an independent contractor to design and build their apps.

Radio Apps are a good way to connect and engage with listeners, making it easy to tune in, have your brand logo show on their devices, or become a source of music information, adding credibility for the station with listeners. They have been proven to help with discovery and prolong listening.

Broadcast Radio Ltd can organise some amazing Apps with stylish clean design and colours to match the station logo. Their apps also include song artwork, station schedule, push notifications and a 'contact our studio' facility. Submission fees due to Apple's iStore and that for Google Play plus the maintenance costs are included in BRL's charges. A discount is available for readers of this book by quoting ORB24 – page 172 has details.

We recommend using app-builders who understand radio, such as BRL , it's always better to work with an experienced broadcast consultant.

ONLINE Radio

Potential Audience for Online Radio

Theoretically, anyone with the necessary equipment could be a potential listener to every radio station using the internet to broadcast. Considering the number of SmartPhones alone, that amounts to 7.9 billion prospective pairs of ears that could listen. It's unlikely, due to language differences (though that's not so relevant if the programme is one of music, which is international in appeal) and the listening tastes and habits of listeners. Another factor that diminishes audience is that we spend a third of our time asleep. And listeners have lots of other things, and radio, to spend their time with however online radio has the potential to access huge audiences.

Some radio listeners use other platforms, but the size of the online audience is growing. Data disclosed by the BBC in Autumn 2023 shows that a quarter of Radio 1's audience do their listening online, only marginally less than the 28% to DAB. The online listening levels BBC Radio 2 are broadly similar (20%) while FM accounts for about 40%. Overall, DAB and FM are falling, unlike Online Radio which continues to grow.

ONLINE Radio

Chapter 11

LUMINARIES & LEADERS

The radio business has such an assortment of bodies and organisations involved that it can be quite bewildering for both listeners and operators, to know exactly who does what. We shall try and explain some of those people in the online radio world and the organisations with which they're involved.

Broadcast policy in any country is usually the prerogative of the government and the UK is no different. Laws are enacted by Parliament but the day-to-day operation is in the hands of ministries, or departments. In the case of media and especially broadcasting, this has been the DCMS (**Department for Culture, Media and Sport)** which has been responsible for broadcasting for over thirty years.

The minister responsible is **Lucy Frazer** who took office in 2023. Her DCMS staff consult with all interested stakeholders and then take policy decisions. In 2024, she will be steering the new Media Bill, through Parliament.

ONLINE Radio

OFCOM

The implementation of the DCMS's work however is devolved to OFCOM, an agency which handles licensing and monitoring activities from its HQ at Riverside House in London.

At the moment, OFCOM has no mandate to control Online Radio and is only involved where stations are transmitted on radio frequencies. There are moves however to extend its remit to licence Online Radio stations. OFCOM licences the multiplex operators who provide the DAB transmission facilities. It also licences programme providers, the radio stations. OFCOM's website (*www.ofcom.org.uk*) contains details of all the UK's licensed multiplexes and other broadcast transmitters as well as companies who hold licences to broadcast on radio frequencies.

Christina Squires
OFCOM

Christina Nicolotti Squires recently joined OFCOM's senior management team as Group Director for Broadcasting and Media and is responsible for the authority's radio licensing operations. She leads the team that is responsible for implementing broadcasting policy and regulation. Christina has more than 30 years of experience in broadcasting, and was recently Director of Content at Sky News. Before this, she was Editor at ITN's *5 News* and has held other senior positions at ITV News having begun her career in print and local TV.

The Radio Centre is an industry body representing the large commercial radio operators, acting in three areas:

ADVERTISING   promotes the benefits of radio to advertisers and agencies.

POLICY   provides a collective voice on radio issues to government, politicians and regulators.

CLEARANCE   ensures that radio advertisements comply with the rules and standards set out in the BCAP and the OFCOM codes.

The **Community Media Association** (see page 170) acts on behalf of most community radio stations. There is no organisation representing online or small commercial radio stations.

ONLINE Radio

BROADCAST PIONEERS

There have been many illustrious pioneers of technology over the last hundred years who were at the forefront of invention and development and whose work made the radio business what it is today. To cover each of the important ones and their work would take a huge book of its own. Shown below are just a few of those pioneers still involved in the business today:

ALAN BEECH

The founder of leading broadcast transmission providers COMMTRONIX and the chief broadcast engineer at Radio Caroline, Alan also built low power AM and FM transmitters for many years, many used by RSLs in the UK and in the Netherlands where he has also been responsible for MW transmitters on another ship, the lightship Radio Seagull.

Alan built one of the first SS-DAB multiplex transmitters and was responsible for its installation and launch in Portsmouth. "The trial broadcasts were intended to test and aid the development of a low-cost method of delivering digital radio services," explains Alan. "With a new licensing framework, it offers small stations an affordable route onto DAB."

In 2017, Radio Caroline was one of the leading online radio stations. It obtained a 'community of interest' licence to transmit on 648MW from an ex-BBC site on Orfordness. Alan installed a 2.5 KW transmitter and, helped by a team that included Peter Chicago, he commissioned the first transmissions. During the pandemic, Alan was the key engineer who brought a Harris DX medium wave transmitter over from Holland and rebuilt it to operate at only a sixth its designed power.

Pioneering engineer Alan Beech
Mandy Marton

The success of those small-scale DAB experimental stations enabled OFCOM to offer six new rounds each of approximately 25 licences over the last five years. The latest batch of multiplexes will be offered in 2024 and extend DAB carriage into the smallest towns and areas in most of the populated parts of the UK. Each multiplex can carry up to 30 different stations. The system enables many online stations to attract an audience locally and attract support and volunteers which has helped many get "off the ground".

ONLINE Radio

PAUL CHANTLER

A respected broadcast consultant, providing valuable strategic, operational and training support. He has helped dozens of stations to launch and is an accomplished author – his book "Hang The DJ" provides essential help for presenters to avoid legal pitfalls.

Paul works in a variety of formats including Adult Contemporary, CHR, Hot AC, Dance, Christian, Sport and Newstalk. Paul has a wealth of experience in radio presentation, production and programme management. In the 1990s, he was a Group Programme Director for three of the UK's biggest radio groups. Paul was responsible for launching two UK regional youth-format stations: *Galaxy* and *Vibe*, as well as helping to launch Ireland's speech station *NewsTalk* in 2002. He also ran UK national station *TalkSport* as Programme Director.

Paul is co-author of a highly-acclaimed and popular textbook on radio journalism, originally published 20 years ago, which has been translated into four languages and is in use in colleges throughout the world. He has also co-written two books on media law. Paul's valuable legal guide for broadcasters **Essential Media Law,** has been updated and is available now in either softback or Kindle form: *https://amzn.to/3SzyyYq*

PAUL BOON

A stalwart radio campaigner, broadcaster, editor and regulator who began his career as a newsreader on Radio Jackie in 1983. Paul chaired the Association for Broadcasting Development (ABD) a lobby group that successfully campaigned for more opportunities for new independent commercial and community radio stations. Paul was architect of the incremental stations idea before joining *The Radio Magazine*, where he became Managing Editor. In 2008 Paul joined OFCOM for nine years working on all kinds of projects, as well as dealing with regulatory matters.

ASH ELFORD

A pioneering multiplex operator, responsible for launching several stations in the Thames Valley. After being responsible for almost two dozen winning applications for licences, Ash had to let the multiplex that he founded be closed but he's now involved with Nation Broadcasting, looking after the *Angel, Your Harrogate* and some of Nation's multiplexes.

ONLINE Radio

LAWRENCE HALLETT

Dr Hallet has over 30 years of experience in UK broadcasting. A founder member of the Community Media Association and director of many broadcast companies, he has advised many successful licence applicants. He has helped to launch community services and commercial stations in the UK and elsewhere in Europe. Between 2004 and 2012, Lawrie was an associate in OFCOM's Radio Team, with responsibility for radio licensing and DAB development.

IAN HICKLING is a multi-lingual broadcast engineer who was an RAF flying officer and a lab technician for J-Beam Aerials. His company ***Transplan UK*** designs, plans and installs radio stations. Ian's mission in life is to enable small radio stations to extend their coverage and to explore and promote all aspects of radio specialising in transmission and propagation. He is a director at online station *Revolution Radio*.

QUENTIN HOWARD

Quentin became the Chief Engineer at the GWR Group where he worked tirelessly to see the adoption of DAB and was appointed CEO of the UK's first commercial multiplex, *Digital One*. Quentin was elected President of World DAB, the international promotional body and then served as Director of Strategy for BFBS, the British Forces Broadcasting Service, at their headquarters. He  modernised the BFBS network and studio centre and ran its *Forces TV* as a commercial TV channel.

Quentin is now an independent media consultant as well as podcaster. Quentin was for many years one of the biggest visionaries in radio, predicting that the medium had a golden future, at a time when others were forecasting its demise. Quentin created Classic *fM*'s birdsong tests which are now available as a podcast. He was the first UK broadcaster to host live radio from home, with a music and phone-in show on Classic *fM*

SAMUEL HUNT

A Leicester-based RF expert who has written many papers on transmission. Since graduating from Lawrence Sheriff in 2003, he has become a partner in **MAXXWAVE** a professional communications company that's the UK's largest independent radio site operator. They provide transmission services for several large stations and also lease out space on masts.

maxxwave.co.uk

ONLINE Radio

STEVE MARSHALL
A well-revered radio name who has been involved in several radio projects in three European countries, including one of the first online stations fifteen years ago with Keith York. He and Gerard Doherty now operate a broadcast consultancy and *River Radio*, a Top 40/CHR station in Northern Ireland.

Email Steve at: *riverradio@gmx.com*

JAMES MULVANY
Manchester-based serial entrepreneur responsible for launching more than five internet companies, including Radio.co, which teaches radio techniques, launches and provides infrastructure for online radio stations.

RASH MUSTAPHA
One of the pioneer engineers wgo took advantage of the improvements in computing technology to replace the expensive hardware of DAB with open source and freely-downloadable software running on everyday computers. What had previously taken up to £10,000 for a studio encoder could now be done with a £30 raspberry pi computer, the size of a credit card.

Rash built the UK's first small-scale DAB 'multiplex' in 2012 and, using an OFCOM 'Test & Development' licence, he built the UK's first small scale DAB 'multiplex'. Rash installed the low power transmitter on the roof of 'Sussex Heights,' the tallest tower block in Brighton. It was driven by cheaply-sourced items of equipment, rather than using the expensive 'professional' gear used by the early DAB multiplexes.

SS-DAB pioneer Rash Mustapha
Liam Mustapha

Rash's 'lower cost technology' experiments were a success, proving that good quality transmissions could be made from equipment that didn't cost "an arm and a leg", as had been the case previously with DAB. They also proved the efficacy of the newer DAB+ standard, which had not previously been used in the UK.

ONLINE Radio

ANDY J LINTON
After six years as a technical support manager at BW Broadcast, Andy set up his own consultancy in 2000. **Total Broadcast Consultants** designs, installs and commissions radio stations, from studios to transmission networks. Andy is an expert in processing and based in Valencia, Spain.

totalbroadcast.net

JOHN ROSBOROUGH
Decades of experience in Northern Ireland broadcasting, with management experience in the technical, editorial, presentation and business fields. John was one of the first PDs at *Downtown Radio* who has worked relentlessly to build an outlet for independent stations in Ulster and founded Belfast DAB, only to see the multiplex awarded to a community not for profit group. He is now focussed on a mux to cover Larne, Carrickfergus and Newtownabbey.

MIKE SPENSER
An American pioneer based in London who ran his Rock & Roll station on a ship in the Baltic before moving it to London as *Radio Trashcan*. Lack of advertiser support has reduced the station's activities but it's adhered to its roots and is one of the longest running online radio stations still playing free independent Rock & Roll music.   *TrashcanRadio.com*.

GARRY STEVENS
Another bright engineer who began as a tower block pirate in the seventies, and then launched one of the first Bulletin Boards for sharing broadcast information. As the cost of broadband fell and more listeners got online, Garry launched multiple online radio channels as tribute stations to Radio North Sea, etc, which continue as some of the first online only radio stations.

BLAKE WILLIAMS
A young American radio engineer recruited for the Laser project, Blake first experimented with online radio in the mid-1990s and enthused the book's author to include it in other radio projects he was involved in. Few radio investors at that time wanted to invest in a medium that was tied to heavy computers and used 'dial up' connections down a phone line, but some papers he wrote have since proved to be totally correct. Blake envisaged that the cost of both computer processing and mobile data would plummet, making many thousands of radio stations economically viable in every area. Both Blakes dreams have come to fruition, making him one of the leding visionaries and one of the online radio pioneers.

ONLINE Radio

Radioplayer

An online radio distribution platform owned by some UK radio companies that offers stations an internet radio web tuner, a set of mobile phone apps and integrations with other devices. Its owners include the BBC, Global, Bauer and the RadioCentre, although the BBC now urge listeners use their Sounds app instead. At the CMA conference in November 2023, Radioplayer announced they would soon offer carriage to community stations. Online Radio stations are however excluded unless their owners have had another station on the player for over a year and they hold an OFCOM licence.

Community Media Association

Founded in 1983 as the **Community Radio Association**, it's now known as the Community Media Association to reflect the convergence of digital communications. A non-profit making organisation, it represents the sector to Government, industry and regulatory bodies. CMA membership includes established organisations, aspirant groups and individuals to whom it provides a range of advice, information and support.

The CMA was led for over twenty years by Operations Director, **Bill Best**. Bill caught the community radio 'bug' while volunteering for an RSL in Sheffield. His specialist areas are community broadcasting technologies, online media delivery, open-source software solutions, and social media networks. Bill is an expert on licensing, music copyright and anything technical.

BITSTREAM BROADCAST

A Lancashire-based radio consultancy providing everything from radio consultancy to system integration, at low cost. From a single transmitter to fully connected networks, Bitstream can advise on planning, installation and ongoing maintenance. Their small team of RF engineers is headed by ex-BBC and Arqiva engineer John Bibby.

RADIO TODAY

A publisher of news and comment about UK radio with regular podcasts and audio newsletters covering all forms of radio, including online. They have similar websites covering Podcasting and Jingles.

ONLINE Radio

RADIO LISTENER'S GUIDE

RLG publishes several very useful annual guides: The *Radio Listener's Guide*, the *Television Viewer's Guide* and the *Mobile Phone User's Guide*. They have been providing technology advice for over 20 years and take pride in providing well-researched, independent and up-to-date information and advice for consumers.

The annual **Radio Listener's Guide** contains a detailed analysis of the various broadcast stations and reception apparatus available to listeners in the UK market as well as giving a good appraisal of the trends in the radio business. The RLG also includes a comprehensive listing of all the radio transmitters in the UK.

The Radio Listener's Guide is the UK's most detailed, comprehensive and independent guide to radio receivers, from portables up to the top of the range 'table top' and Hi-Fi radios. These honest reviews are an invaluable guide for intending buyers plus the RLG also has latest news for listeners. You can order Radio Listener Guide direct at: *www.radioguide.co.uk*

RADIO MONITOR
A company that monitors all the music played on radio and TV in over 120 countries. Their airplay data is used by major record labels, media and for royalty management. Access to this data online tools, via the Radiomonitor App, by emailed reports or through a custom data feed.

MEDIA SHOW
A weekly programme on BBC Radio 4 about all media, including social, anti-social and news. Very occasionally, the programme mentions online radio. New episodes are released every Wednesday with previous editions being available to download on BBC Sounds.

All the series' most recent programmes can be heard via the BBC Sounds app and the producers of can be contacted by email at: themediashow@BBC.co.uk.

ONLINE Radio

Podcast Radio uses an online channel to showcase a variety of podcasts to the world, combining live presenters, news updates and charts. The Podcast Radio Network covers the entire digital audio ecosystem; pod-fans can now discover podcasts on their radios via DAB+ in London, Surrey, Manchester and Glasgow. Podcast Radio is run by Gerry Edwards and Paul Chantler, the veteran radio consultant who has been a midwife at dozens of innovative radio station launches. The channel has now expanded into North America. An excellent example of how linear online radio can present the best of 'on demand' programming, or podcasting as it's now known.

Worldwide Broadcast Consultants

An international consortium of broadcast experts formed in 1983 by a team of English and American engineers. They are highly skilled in licensing, administration, engineering, programming and sales techniques to establish radio stations, whatever the mode of delivery.

Led by the author of this book, WBC acts as agent for consultants in several countries and has completed work for both regulators and radio station operators. The WBC team have many years of experience building successful radio stations of all sizes and in over 24 countries. The WBC team respect investors' desire for secrecy and undertake all engagements with the utmost confidentiality under the protection of suitable Non-Disclosure Agreements.

Among their services provided are:
CONCEPTION identify the market
LICENCE Identify and obtain licences.
DESIGN & BUILD the infrastructure.
FORMAT refine and develop niche formats.
LAUNCH carefully plan and supervise for maximum impact.
MARKET the station and its audience to advertisers.
CURATE monitor and train team, developing new revenue streams.

https://bit.ly/3SiRg9W

Chapter 12

SUPPLIERS

BROADCAST RADIO
A company run by a team of expert staff with decades of experience both designing solutions and working in real radio stations. Founders Peter Jarrett and Liam Burke developed one of the first radio station software solutions while at Hull University in the early 1990s - **P-Squared**. It has remained the cornerstone of the playout system at many radio stations since its launch in 1997. Myriad can now be found in over a thousand stations, from nationals right down to the smallest community outlets, who have all found its intuitive software among the easiest to learn.

The Yorkshire-based company has now expanded into many branches of broadcasting, meaning they can help guide you through both the technology and the creative process. *Broadcast Radio* offer bespoke complete 'turnkey' radio stations: studio design and installation, software training, streaming and hosting services. New stations are offered some amazing deals, such as ways to start broadcasting immediately using **Myriad Cloud Radio,** the world's first professional grade radio software, designed for 24/7 operation, natively in the Microsoft Azure cloud. (Quote Ref ORB-24 for a discount).

www.broadcastradio.com

ONLINE Radio

Broadcast Radio have a wide range of professional studio and distribution equipment and services to help broadcasters launch and stay on the air. For over 25 years, the boys at BR have been providing the radio industry with top-notch radio software, such as Myriad, loved by staff at thousands of radio stations. The Myriad 5 Playout system can now be accessed remotely by the Cloud by adding **Myriad Anywhere**, enabling use of a web browser to edit and voice-track audio.

Myriad 6, was redesigned from the ground up in 2023 to provide the ultimate connected experience for presenters and enable them to make great radio. Suitable for stations of all types, it can be operated as 'live-assist, full automated, as an audio production facility or for remote VT. It has *SmartInfo* that dynamically aggregates relevant artist details and other information that makes any presenter sound like an expert as well as clock and rule scheduling.

BR's software is used at thousands of stations around the world and they recently expanded into the USA. **Myriad Cloud** is the world's first professional radio software designed for full-time use that runs natively on the Microsoft's Azure cloud platform. *Myriad Cloud Radio* employs the same cutting-edge technology found in the latest *Myriad 6* software, giving unrivalled functionality without the need for physical studios, local software, or infrastructure. No VM's, no studio, no virtual audio devices; it's all browser based, so stations can be managed from any PC. It's available in various tiers, from £70 a month.

Myriad News is a hybrid newsroom system for modern journalism that can capture and manage news feeds, importing content from subscription sources such as IRN / Sky News & the PA as well as general sources such as RSS feeds. Combining this with locally generated content it compiles bulletins and records news breaks for uploading and broadcast.

Broadcast Radio build and supply complete radio stations, supplying all the studio equipment including desking, mixing consoles, playout systems, microphones, codecs and processors. Among the many support services supplied by BRL are station web sites, **Mobile Apps** and **Alexa** skills.

Readers can get a valuable discount by quoting the reference ORB-24.
Call them on 01482 35070.   *www.BroadcastRadio.com*

ONLINE Radio

AIIR
An international software supplier based in New York, Canada and the UK, they offer their own *PlayOutONE* automation system and a cloud-based scheduler as well as *Studio Inbox* combining listener social media interaction, mobile apps and Alexa skills.   Aiir.com

Amazon
The biggest online retailer in the world, Amazon stock a wide range of DAB Radios, plus studio equipment and other items that listeners and broadcasters will find useful. The field of broadcast is an unusual one where many items are not available close by, especially for stations in more remote and areas. Amazon usually have huge stocks of key items and the accessories found in the broadcast world and most are available for 'next day' or even 'same day' delivery, depending on destination. This includes microphones, mixing consoles, playout equipment, headphones, etc. The new *RodecasterPro* is a good example; now used by hundreds of broadcasters to make radio programmes at home.

https://amzn.to/3LLxRsX

AUDIO BROADCAST CONSULTANTS
A small, friendly bunch of experienced radio professionals offering independent advice. They don't sell kit and so they are not influenced by sales commissions when they make a recommendation. ABC specialise in FM radio transmission, but also have wide experience across the broadcast industry from studio specification to station imaging. The ABC expertise comes from over 20 years' experience, including using computerised radio propagation modelling tools to quickly and accurately deliver radio coverage predictions / maps / plots.

ABC is led by **Glyn Roylance** who has many years of broadcast expertise and have been involved in exporting equipment too. Glyn was a Broadcast Engineer at the IBA, before moving to Arqiva, where he was responsible for installing, commissioning, operating and maintaining TV and radio networks such as Capital, LBC, BFBS, Channel 4 and ITV.

His brother Clive is also part of ABC; he embarked on a radio career at the Voice of Peace "Somewhere in the Mediterranean". He set up and ran two stations in Ireland for many years as well as being heard as a voiceover on several UK stations.

ONLINE Radio

COMMTRONIX

A leading DAB transmission consultancy, supplier and installer of everything needed to get a DAB station on air, Commtronix was founded by two former NTL and Arqiva engineers over ten years ago and has played a pivotal role in SSDAB since the launch in 2015. Firstly, with the Portsmouth trial, then Glasgow and then supplying equipment and knowledge to almost all of the other SSDAB operators.

Commtronix are one of the few UK companies involved with SSDAB to hold VIA licence patent agreement to distribute the patented AAC+ coding software used for DAB+ and collect the associated royalties. They were the first to migrate SSDAB encoding from the OFCOM supplied systems onto lower cost raspberry pi computers.

Offering a wide range of services for the broadcast and telecoms industry, from concept to commissioning, they cover the whole of the British Isles. Commtronix provide an end-to-end solution based on their 25 years of on-site practical experience of making radio happen. They offer transmission and multiplex install and maintenance to digital radio operators, as well as conventional AM and FM broadcasters.

Commtronix founders **Alan Beech** and **Wallace McKeown** each have experience of writing and implementing software test routines, debugging prototype commercial software, component level. They have worked on all manner of broadcast transmission and communications equipment. There is not much in the world of communications and electronics engineering that they've not been involved with.

http://www.commtronix.co.uk

SKY NEWS

Sky News supplies the national and international news, sport, business and entertainment news to almost every commercial radio station in the UK. The dedicated team of journalists work from the Sky News Centre near Heathrow; The copy and audio are sent directly to radio stations own newsrooms for live reading to air, or stations can choose to opt in to the hourly bulletin. Sky News was pioneered by Andy Ivy in the 1990s as a competitor to the IRN service.

https://news.sky.com/info/radio

ONLINE Radio

DIGITAL RADIO CHOICE
Digital Radio Choice is a buyer's guide for digital radios in the UK. Its website provides information about digital radios, answers to common questions and updates on programmes and stations. The aim of their website is to help buyers find the best new radios, etc, and it is funded by affiliate commissions they earn by referring buyers to radio retailers.
https://www.digitalradiochoice.com

EBUYER
One of the largest online retailers of electronic components and equipment, they have over four million registered buyers and 250 staff. Its purpose-built hub is just a few hundred yards from the M62 motorway, centrally located and ideal to ship to any UK or EU destination immediately. They hold vast stocks of computer equipment, peripherals and components. Ebuyer's philosophy is simple: "give customers what they want, when they want it."
Ebuyer's latest clearance lines: *https://tidd.ly/36En4wL*

LUCORO
A well-respected builder of transmitters and other radio equipment for over 25 years, they source and build all their equipment near Harrogate in Yorkshire to the full specifications demanded by OFCOM and provide second-to-none after care service.
https://www.lucorobroadcast.com

RADIO STRUCTURES
A Northampton-based fabricator of masts and substructures since 1979. RS can also provide a rigging and installation service, from site surveys through to final installation. They also run training courses for mast climbers, to the strictest Arqiva safety standards.   *www.radiostructures.com*

RADIO.CO
A Manchester-based company founded in 2015 by **James Mulvany**. Specialising in online radio, they have helped many stations to launch with their own architecture, built completely from the ground up. James Mulvany, has been at the forefront of the online radio industry for the past decade and has a large talented team of IT and programming experts. *Radio.co* guide operators, train staff and offer cloud-based systems to make it simple to launch small radio stations, online or over the air.

ONLINE Radio

Chapter 13
GLOSSARY

AAC Advanced Audio Coding is a system of digital used in DAB+ and audio streaming. A widely used standard and was designed to succeed MP3.

Aggregator A database of links to streams, used by a radio to find the various station feeds.

AIIR Broadcast software company producing automation, studio message centre and Apps for radio stations.

AM Amplitude Modulation. Explained in Chapter1.

Analogue Audio in its natural form, with undulating waveforms. In, contrast, digital is audio chopped up into segments with only some surviving.

Apps Applications (often called widgets) to ease quick tuning of radio stations on a mobile, but also found on desktop PCs.

BANDS The spectrum is divided into bands of frequencies. The best known are Medium Wave, Short Wave and the VHF broadcast Band II (88 to 108 MHz). DAB is found in VHF Band III (174 to 240 MHz)

BER Bit Error Rate is calculated by comparing the transmitted of bits to the received bits and counting the number of errors. The ratio of how many bits received in error over the number of total bits received is the BER.

Catch Up listening to radio after the live transmission ends, usually 'on demand'.

Comscore The currency used in planning and buying media time.

DAB Digital Audio Broadcasting, first transmitted in 1988.

DAB+ A newer DAB encoding system which makes more efficient use of the spectrum and squeezes more stations through the same transmitter at similar quality.

DSIT Was originally the DCMS (Digital, Culture, Media and Sport) ministry of the UK Government before being reshuffled to Dep't for Science, Innovation and Technology. They are ultimately responsible for all broadcasting and they authorise OFCOM to manage the spectrum and licence radio transmission.

ONLINE Radio

De-Esser Processor that reduces gain to suppress sibilance and other high frequency issues in a vocal track.

DMB Digital Multimedia Broadcasting - sometimes referred to as mobile TV. It's a sibling of DAB (Digital Audio Broadcasting) and DAB+.

DRM Digital Radio Mondial, a system for adding digital signals in bands for analogue, e.g. MW, SW and FM

DSO Digital Switch Over, a date when analogue transmissions cease and digital becomes the sole transmission method.

DX Very long distant radio reception, out of the normal service area.

DXers A radio enthusiast who listens to rare distant radio stations.

EBU European Broadcasting Union

EECC European Electronic Communications Code An EU directive that makes it mandatory to install a digital radio receiver in every new car. This can be DAB, DRM or any digital broadcast received online.

Ensemble A group or a collection of radio stations. The word ensemble is often used interchangeably with multiplex.

EPG Electronic Programme Guide, an over the air menu, used on TV for 35 years now and found on some digital radios.

FM Frequency Modulation – an analogue method of adding programme or data to a radio carrier.

FIC. Fast Information Channel. A term used for signals with low data rates; e.g. paging, traffic info, or messaging.

FSK. Frequency Shift Keying. Adding data to a signal digitally.

GHz Giga Hertz. One thousand million hertz. (10^9)

GSM Global System for mobile communications.
HD Radio In-band on-channel, usually called IBOC, is a hybrid process of transmitting both analogue and digital signals together on a single frequency. Used on many AM channels in North America.

ONLINE Radio

HE-AAC High Efficiency Advanced Audio Coding.

IAB The Internet Advertising Bureau oversees digital advertising for its 1,200 members including media owners, agencies and brands.

IBOC see HD Radio

ICE In Car entertainment, which includes all kinds of audio equipment plus video and gaming too.

IP (a) Internet Protocol. A unique address used by computers to route data
(b) Intellectual Property. An invention or patented or copyrighted work.

IPDTL Internet protocol 'down the line', using browser for live remotes.

IRT Institute fur Rundfunk Technik (the Institute for Broadcast Technology) in Munich that invented DAB.

ISDN Integrated Services Digital Network - a now obsolete standard of transmitting audio and data along phone lines.

ITU *International Telecommunications Union* – Based in Geneva this is a global organisation whose membership is mainly communications and broadcast regulators from almost every nation in the world. It sets standards and agrees usage of the electro-magnetic spectrum.

JICRAR The body that collated radio listening figures for commercial radio until 1992; now taken over by RAJAR.

LSF Low Sampling Frequency. Encoding with halved sampling rate.

LW Long Wave is a band of wavelengths from 2000m to 1000m (from 153 KHz to 279 KHz) used for broadcasting since the 1920s in some parts of the world. They travel very long distances and are usually very high power.

M/bits Megabits (10^6) per second, usually written mbps

MHz Megahertz, one million (10^6) Hertz.

MER Modulation Error Rate provides an indication on how well each symbol was received, often used in transmitter testing. MER is defined as the ratio of average symbol power to average error power. The optimum value should be at least 20dB, and never lower than 10dB.

ONLINE Radio

MiniMux common term for a small multiplex (often applied to SS DAB).

MP3 MPEG-1 Audio Layer III or MPEG-2 Audio Layer III - a coding format for digital audio. Other coding formats include MP3, AAC, Vorbis, FLAC and Opus. (see page 53 for a comparison of file types)

MPEG Moving Picture Experts Group. A working party of authorities in the audio, video, TV and radio industries to set standards for digital compression.

Multiplex a word describing the transmission of multiple radio stations or audio channels over a frequency.

MUX An abbreviation of the word multiplex.

MUXCO Muxco is a UK consortium which operates regional DAB multiplexes

MW Medium Wave is band of wavelengths from 550m to 188m (corresponding to frequencies from 531 to 1602 KHz) that have been traditionally used for AM broadcasting.

NIELSEN US company focussed on media monitoring.

OFCOM The UK regulator, authorised to manage communications, including radio broadcasting. They licence all radio transmissions in the UK.

OFDM Orthogonal Frequency Division Multiplexing is a type of digital transmissions, encoding data or programming onto several carriers, used in wideband transmission.

Podcast a digital audio programme, usually one of a series, that can be downloaded 'on demand' by listeners.

PDM Pulse Duration Modulation a high efficiency technique used in high power AM transmission. A variation of PCM was called Pulse Step Modulation.

PSK Phase Shift Keying, a digital modulation process used for data.

QAM Quadrature Amplitude Modulation is several methods of applying digital modulation to a carrier. Two feeds are added to the carrier, out of phase.

RDS Radio Data System. Adding data to a broadcast signal. Used since 1995 for station identification, time signals and information onto FM.

ONLINE Radio

RAB The Radio Advertising Bureau represents the US broadcast industry, primarily with resources to attract new sales and promote professionalism by training.

RAJAR UK radio audience research organisation.

RF Radio Frequency. Usually applied to an electro-magnetic wave (radio energy). From as low as 16KHz up to many GHz

Satellite An orbiting spacecraft, that relays signals giving a very wide area coverage.

SDL Sound Digital Ltd is the operator of the UK's second 'national' network, of commercial DAB stations which reaches 83% of the population.

SDR Software Defined Radio – a complete 'over the air' radio receiver 'on a stick', plugged into a computer's USB socket.

SFN Single Frequency Network.

Short Wave Also known as HF (High frequency). Uses frequencies from 3 to 30MHz, which propagate by 'skipping' between earth and the various tropospheric levels around it.

SMMT A trade body for the UK automotive industry and *In Car Entertainment*.

Streaming sending programmes by internet

TotallyRadio the UK's first multi-channel radio aggregator.

Webcast Media presentations streamed over the Internet from a single source to many simultaneous listeners or viewers.

ONLINE Radio

The Future is Online Radio

Ever since Baird first demonstrated television in the 1920s pessimists have predicted the end of radio. In recent decades, it appeared as though they may be right, when DAB's prohibitively high prices and other difficulties seemed to be the final nails in radio's coffin. A combination of over-regulation, heartless management and zealous accountants have stifled radio's creatives. Radio seemed to be gasping its last breath as the energy and enthusiasm were drained out of the medium, so loved by around 90% of the population for almost a hundred years.

Radio thrives on personality and on the relevance to its listeners. Success often follows when there are opportunities for local people to influence their local stations, which are the grass roots where most national broadcasters take their first steps and learn their skills. Much more than a simple jukebox, means so many things to so many people. Personality is the dominant factor; many listeners regard their favourite stations as their friend. Its service as companion meets many needs and long may it continue to do so!

Some broadcasters have excluded newcomers from what they regard as their own personal fiefdom; for the last two decades they demanded that all AM and FM be shut down, to give DAB a clear run at becoming the only type of sound broadcasting. They've failed so far and analogue continues. DAB's lunch meanwhile is being nibbled away by new independent broadcasters audible on small-scale multiplexes, which is fine for neighbourhood or local radio but Online is special; it can be global.

While the transmission medium is not important as it's the content that is the message. TV and radio audiences are best served ONLINE, where the audio fidelity and the market size are unlimited. When numbers matter, there are almost 7 BILLION smartphones now in use globally and the only radio platform that most can access is ONLINE. That is surely the future?

Stay tuned!

Paul Rusling.
November 2023

ONLINE Radio

More Information

More information about radio station equipment can be found on the World of Radio web site's **TRANSMISSION** section, which has separate pages for mixing desks, audio processors, microphones, recoding and library archives (HD and SSD) and all kinds of equipment needed to equip a radio station, over the air or online. The section also has advice on hosts (internet providers) and playout automation.

https://WorldofRadio.co.uk/Transmission.html

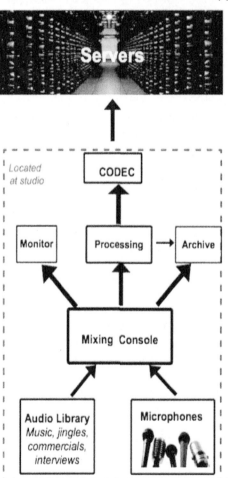

The Online Radio audio chain can appear daunting at first but, after reading this book, it will hopefully by now seem very logical? TV is already digital and their preferred transmission method is going online.

It's important to know what each link in the radio station's audio chain actually does. Sometimes, the only solution is to either do it personally, or get the advice of someone who has already done so and can identify the pitfalls, pointing out the best and the most economic direction.

After working as a consultant to several radio broadcasters and regulators, I know that while professional fees MAY initially seem to be expensive, using an amateur can often cost even more.

YOUR thoughts or comments would be very welcome, as would your questions. What do YOU want to know? Do let me know. Please do email me:
Paul@OnlineRadioBook.com

ONLINE Radio

RADIO FORMATS
in the UK and US

A useful book describing the evolution of radio station formats on both sides of the pond over the last hundred years.

Radio Formats in the UK and US tells how speech and music radio in the USA followed public demand into sharply defined streams, following music genres. It also describes how the radio landscape in the UK was stifled by regulators. Over 130 radio formats are described in the book's 100 pages. It guides readers through radio's halcyon days, highlighting some of the milestones of broadcasting and introduces the pioneers who developed particular formats. The UK's BBC stations (national, regional and local) are covered as are commercial stations, community radio and both onshore and offshore pirates.

Available as a Paperback or Kindle via Amazon: https://amzn.to/3SvVNTI

For a range of books about the radio industry, visit
https://WorldofRadio.co.uk/RadioBooks.html

The Online Radio Book has a new service:
Regular **up-to-date news** about Online Radio
Plus suggestions of fresh
Revenue Generation ideas
for online radio stations
More details:
OnlineRadioBook.com

ONLINE Radio

🔔 Radio Caroline bible

The full story of how the world's most famous offshore radio ship attracted millions of listeners but finally came ashore and is a part of the UK's burgeoning DAB Network. Radio Caroline is now the longest running station using the DAB+ standard.

The RADIO CAROLINE BIBLE tells the story of the station's life and explains many previously untold aspects from the point of view of those who were there.

With input from key Radio Caroline team members, the book tells chronologically what really happened around the start of the project, the fund-raising process, who put up the money, why they did so and what was the legal basis for the station.

The full story of Radio Caroline's sixties operation, the difficulties that led to the ships being towed away and what happened in Holland. Finally, the mystery man who paid to set the Mi Amigo free once again is revealed also ng with the story behind those lucrative contracts to transmit Dutch stations, the Caroline Homes and those dastardly government raids.

The story is brought right up to date with Caroline's Chinese connection, the motor racing links, the relaunch using the former BBC World Service site and channel plus Caroline Flashback, the new community Caroline station in Essex and another new boat! This hardback book is indexed and illustrated with over 300 black and white photographs; a story that will keep you amused and amazed for hours.

See the book's own website for fuller details:
https://RadioCarolineBible.com

ONLINE Radio

Podcasting is the newest media and it's growing at a phenomenal rate. While downloading audio began in the 1990s, it was only in 2004 that the word **podcast** was first coined and it took ten years to get any real traction.

In recent years podcasting has snowballed and the delivery of individual episodes of radio programmes "on demand" has attracted a lot of radio's listeners.

Is podcasting simply "radio on demand", with the listeners deciding the schedule? This book describes what a podcast is and how it's made. It guides the reader through the history of the medium and the equipment needed to produce a podcast. It looks at many of the 'movers and shakers' in the industry, the corporate players and sources of media coverage about this latest way to disseminate or obtain education, news and entertainment.

Author Paul Rusling draws on his decades in radio broadcasting and podcast consulting to steer readers through the minefield of studio equipment, web host servers, directories, distribution and audience metrics, to give a broad overview of this exciting new medium. This is a hundred-page book, available immediately in Softback or Kindle.

https://worldofradio.co.uk/Podcasting.html

ONLINE Radio

More radio books

Radio Adventures of the MV Communicator
200 pages of swash-buckling fun on the high seas. The excitement, the action and real piracy as eleven stations transmitted by one small ship on the North Sea.

Laser Radio Programming
How a new radio ship brought innovation to British radio and won over ten million listeners. Details of all the station's personalities plus many format secrets and her operating manual.

Internet Radio 2016
written in 2016 and describes all the equipment then needed to broadcast on the internet. The book also describes the processes for getting an online radio station heard on networks around the world.

DAB & DAB+. The future of radio?
describes the development of DAB and explains the growth of the new Small Scale DAB stations.

Radio Caroline Voices on the Air
Lists all the voices heard on Radio Caroline over the last six decades. DJs, newsreaders, etc and discusses DJ hirings and formats.

Podcasting ~ a guide
Describes podcasts, how they are made, and guides readers through the medium's history and the technical aspects of the equipment. Examines the kit, the distributors, 'movers and shakers' in the industry, the corporate players, etc, in this latest medium – often described as 'radio on demand'?

All available as softback books or as Kindles
https://worldofradio.co.uk/RadioBooks.html

ONLINE Radio

INDEX

Acoustics	136 - 138	Codecs	140
Affiliate sales	125, 126	Combs, Scott	26
AIFF	69	Comm Media Association	166
AKG mics	84, 85	Commtronix	171
Alexa Skills	157	Compression	45
Amazon Echo Show	156, 170	Copyright	54, 55, 56, 61
Apps	158	Cridland, James	4
Arakis	100	D&R	97, 101
Atlantis FM	40	DAB, start & demise	24, 40
Audio Broadcast Cons	170	DAT	71
Audio HiJack	141, 143	Digital Workstations	38
Audio storage	69, 70	Day, Rob	40
Audio Technics Mics	85	Delicast	145
AudioRealm	145	Digital audio files	45
Axia	100	Digital Radio Choice	172
Beech, Alan	4, 162, 171	Digital TV	42
Behringer	90, 99	Distributors	139-148
Best, Bill	166	DMCA	26
BIT rates	44	DRM	41, 48
Bitstream Broadcast	166	DSIT	159
Boon, Paul	163	Ebuyer	172
Branding	109	Elford, Ash	165
Broadcast Radio Ltd	169	EM Spectrum	30
Broadcast.com	27	Evans, Chris	68
BUTT	141	Finance overheads	123, 127
Cambridge Majority	155	Financial Revenues	123, 128
Car radios	157	FM, the birth	21, 22
CDs	72	Frazer, Lucy, MP, Minister	163
Chantler, Paul	53, 163	FreeSat	50
Classic FM	24, 44, 162	Frequency	12

ONLINE Radio

Funding	124	MCPS	58
Future of Online Radio	128, 179	Media Show	167
Glossary	174	Mic' Lavalier	80
Hallett, Dr Lawrence	164	Mic' patterns	79
HDRadio	39, 42	Microphones	73 - 88
Headphones	93 - 96	Mixing desk / consoles	97
Hickling, Ian	164	Modulation	33
Howard, Quentin	24, 164	Music Royalties	54, 55, 56
Hunt, Sam	165	Mustapha, Rash	161
Hyde, Chris	109	Mulvany, James	144, 167
IBC	14	MyTuner	146
Ibiquity	42	NetRadio	26
Icecast	141	Neumann	83
ipDTL	67, 68	Normandie, Radio	14
Jamming	32	Numark	94
Jingles	134 - 135	OFCOM	60, 160
Kiss FM	23	Omnia	91, 92
KPMX, KSAN	22, 23	Online Radio Directories	145
Kramer, Kane	16	Operations	129
Lawo	102	Plugge, Leonard	150
Leach, Kevin	68	Podcast Radio	168
Lemega radios	153	Podcasting	181
Licensing	52, 53	PPL	56, 57
Linton, Andy	165	Processors, Mic	88, 89, 90
Listen2MyRadio	142	Processors, Output	91
Lossless files	45	Programme Content	53
Luxembourg, Radio	15	Programme Formats	105 - 122
LW	46	Programming	133
Malmud, Carl	16	Promotion	127
Marconi	20	PRS	58
Marketing	127	Pure Rock Radio	25
Marshall, Steve	165	Radio Books	182, 183

ONLINE Radio

Radio Listener Guide	155, 167	Shure Mics	78
Radio Monitor	167	Simple Radio	147,158
Radio Structures	172	SiriusXM	43
Radio Today	166	Sky News	171
Radio waves	11, 29,	Small Scale DAV	49
Radio.co	144, 172	Smart Speakers	157
RadioDJ	143	Sonifex	103
RadioFeeds	146	Sound of Spitfire	109
RadioGuide	146	Spenser, Mike	167
RadioLogik	143	Sponsorship	124
Radionomy	146	Squires, Christina	160
RadioPlayer	166	Station Names	107-108
RadioStation PRO	140	StreamPlayerPRO	140
RCB	182	Streema	147
Receivers	149-159	Studio equipment	63-104
Recording	36	Tag lines	108
Reeves, Alan	16	Telephone kit	67
Research	130-132	Transmission	139
Revenues	123	TuneIn	147, 158
Roberts	154	UK Radio Portal	42
Røde mics	82	V-Tuner	148
Rosborough, John	165	VHF-FM	47-48
Roylance, Glyn	170	Viamux	172
Ruark	152	Virgin Radio	25, 68
Rusling, Paul	179, 184	VirtualDJ	143
SAM Broadcaster	143	WAV	69
Satellite	43	Wavebands	31
SaveNet Radio	26,27	WBC	168
SDR (software defined)	158	Webcasting	18
Sennheiser	94, 96	Windows Media Radio	148
Servers	140	Worldspace	43
Shoutcheap	142	York, Keith	26

ABOUT THE AUTHOR

Paul Rusling studied radio engineering in the early 1970s, financed by working as a night club DJ. In 1973 he joined *Radio Caroline* as a disc jockey where he hosted the station's breakfast programme. He and his wife Anne managed clubs and pubs in London and the south for some years before working as a broadcast consultant.

In 1983, Paul converted a research ship, the MV Communicator, for use as a floating radio station. The ship broadcast almost a dozen different stations in its 21 years career at sea, but found most success in its days as *Laser 558*, which attracted ten million listeners. It made most money broadcasting as a licensed station in the Netherlands as *Veronica Hit Radio*.

Paul has worked as a broadcast consultant for many large-scale radio stations in several countries. His work has often included programming and engineering and he was a consultant to media regulators and to several state and private organisations, across Europe. His contract include work from Lithuania and The Lebanon in the east, across to Morocco and the Azores in the west. Clients have included the *SBC, NOZEMA, COPE, RTL, Sky Radio, Veronica Omroep, Radio 10, Virgin Radio* and *Classic fM*.

As well as writing on media for magazines and periodicals, Paul has had several books published about radio. Recent books include *Radio Formats in the UK and US, DAB & DAB+* and several books about radio stations, such as Laser Radio programming and the Radio Caroline Bible.
All are available at: *https://worldofradio.co.uk/RadioBooks.html*

Email any enquiries to
Paul@OnlineRadioBook.com

Printed in Great Britain
by Amazon